Metal–Metal Bonded Carbonyl Dimers and Clusters

Catherine E. Housecroft

Lecturer in Inorganic Chemistry at The University of Basel, Switzerland

Series sponsor: **ZENECA**

ZENECA is a major international company active in four main areas of business: Pharmaceuticals, Agrochemicals and Seeds, Specialty Chemicals, and Biological Products.

ZENECA's skill and innovative ideas in organic chemistry and bioscience create products and services which improve the world's health, nutrition, environment, and quality of life.

ZENECA is committed to the support of education in chemistry and chemical engineering.

OXFORD NEW YORK TOKYO
OXFORD UNIVERSITY PRESS
1996

Oxford University Press, Walton Street, Oxford OX2 6DP

Oxford   New York
Athens   Auckland   Bangkok   Bombay
Calcutta   Cape Town   Dar es Salaam   Delhi
Florence   Hong Kong   Istanbul   Karachi
Kuala Lumpur   Madras   Madrid   Melbourne
Mexico City   Nairobi   Paris   Singapore
Taipei   Tokyo   Toronto
and associated companies in
Berlin   Ibadan

Oxford is a trade mark of Oxford University Press

Published in the United States
by Oxford University Press Inc., New York

A catalogue record for this book is available from the British Library

Library of Congress Cataloging in Publication Data
(Data applied for)
ISBN   0 19 855859 7

Typeset by the author
Printed in Great Britain by
The Bath Press, Somerset

Series Editor's Foreword

The carbonyl ligand acts as a supporting group to stabilize a vast variety of complexes, many of which are polynuclear. The dimers and clusters display an elegant structural chemistry, providing both challenges for contemporary theory and a framework on which to build new organometallic chemistry; this can provide excellent models for species on metal surfaces and heterogenous catalysts.

Oxford Chemistry Primers are designed to give a concise introduction to all chemistry students by providing material that would usually be covered in an 8–10 lecture course. As well as providing up-to-date information, this series will provide explanations and rationales that form the framework of the understanding of inorganic chemistry. Following her Primer on *Cluster Molecules of the p-block elements*, Catherine Housecroft has provided us with another detailed, authoritative description of polynuclear compounds. This will give both those planning courses and undergraduates alike many interesting examples of the chemistry of polynuclear metal carbonyls.

John Evans
Department of Chemistry
University of Southampton

Preface

This text is designed in part as a companion volume to the Primer *Cluster molecules of the p-block elements. Metal–metal bonded carbonyl dimers and clusters* deals with the synthesis, structure, bonding, and reactivity of *d*-block metal carbonyl compounds and derivatives containing hydride, phosphine, phosphite, and organic ligands as well as clusters containing both transition metal and main group fragments. Like most areas of chemistry, this is a large one and this text is only intended to be an introduction to the field; the reader is guided to several more advanced texts and literature reviews for more in-depth explorations of metal carbonyl dimer and cluster chemistry. The bonding in clusters has been the focus of much theoretical attention over the years, and this book introduces the basics of cluster electron counting—an adequate coverage to allow the reader to rationalize many of the polyhedral assemblies that metal atoms adopt in molecular carbonyl and derivative species. Whilst we consider the structures and bonding of some high nuclearity clusters, the discussions of reactivity focus mainly on smaller molecules for reasons of (relative) simplicity. The final section of the text provides a number of problems which aim to test knowledge of principles as well as an ability to interpret experimental data.

I should like to thank colleagues for time spent in proof-reading. As always, my thanks go to my husband, Edwin Constable, for his help and comments during writing and for his critical views of cluster chemistry which, although infuriating, help focus my attention on the less rational aspects of the topic. Philby and Isis have, as ever, played more than a small role—to them I dedicate this work.

Basel   C. E. H.
January 1996

Contents

1 Introduction and definitions

1.1 How is a metal–metal bonded dimer defined?

In this book, the word 'dimer' will be used in a strict sense — a molecule is classed as *dimeric* if it is of a form $[\{ML_n\}_2]$ or $[\{ML_nL'_m\}_2]$ (or similar) where M is a metal and L and L' are ligands. The formation of a metal–metal bonding interaction is *not*, however, a prerequisite to dimer formation; for example, halides, peroxides, sulfides, or organic ligands may bridge between the two metal centres and support the dinuclear framework (Fig. 1.1). On the other hand, it is important to note that the presence of bridging ligands does not necessarily preclude the formation of a metal–metal bond as we shall discuss in Chapter 5. In this text we restrict our attention to compounds in which the metal atoms are within bonding contact, and two examples of *metal–metal bonded dimers* are shown in Fig. 1.2.

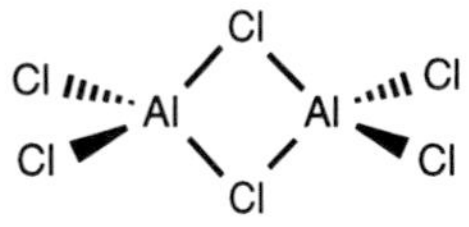

Fig. 1.1   In Al_2Cl_6, there is no direct metal–metal bond.

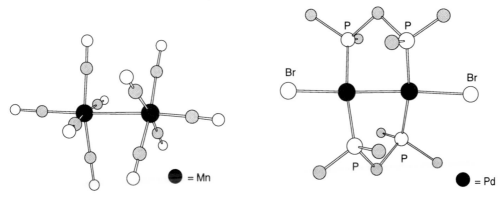

Fig. 1.2   The structures of the dimers $Mn_2(CO)_{10}$ and $Pd_2Br_2(\mu\text{-}Me_2PCH_2PMe_2)_2$; H atoms in the second structure have been omitted.

1.2 How is a metal–metal bonded cluster defined?

The definition of a cluster is ambiguous. In the accompanying text in this series *Cluster molecules of the p-block elements*, we consider a cluster to be a neutral or charged species in which there is a polycyclic array of atoms. This definition excludes monocyclic species such as S_8, but includes organic molecules such as adamantane, $C_{10}H_{16}$, that many chemists may argue is not a cluster at all! When we come to the *d*-block, the term cluster certainly encompasses a wide range of polynuclear metal carbonyl compounds such as $Ir_4(CO)_{12}$ (Fig.1.3) and $Rh_6(CO)_{16}$ as well as polynuclear metal halides such as $[Ta_6Cl_{18}]^{2-}$ (Fig. 1.4) and polyoxoanions exemplified in Fig. 1.5 by

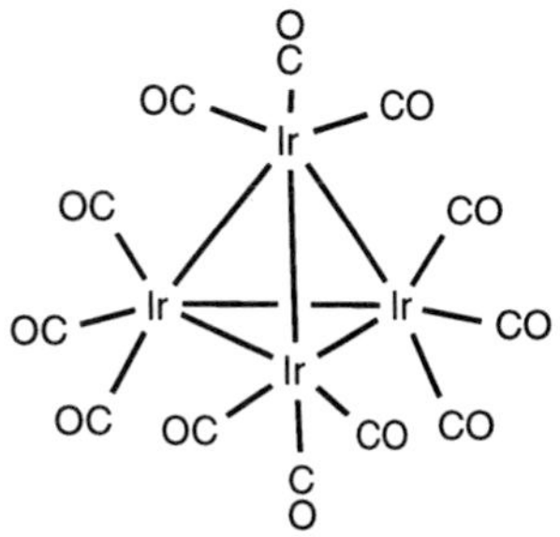

Fig. 1.3   The structure of $Ir_4(CO)_{12}$.

$[Mo_6O_{19}]^{2-}$. In each of the anions $[Ta_6Cl_{18}]^{2-}$ and $[Mo_6O_{19}]^{2-}$, the metal atoms are arranged in an octahedral unit. Figure 1.4 shows the presence of Ta–Ta bonds but Fig. 1.5 indicates that the polyoxoanion has a more open structure which is supported by the oxo-bridges and includes an interstitial oxygen atom.

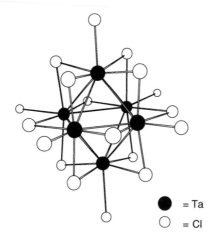

● = Ta	○ = Cl

Fig. 1.4 The structure of the $[Ta_6Cl_{18}]^{2-}$ dianion.

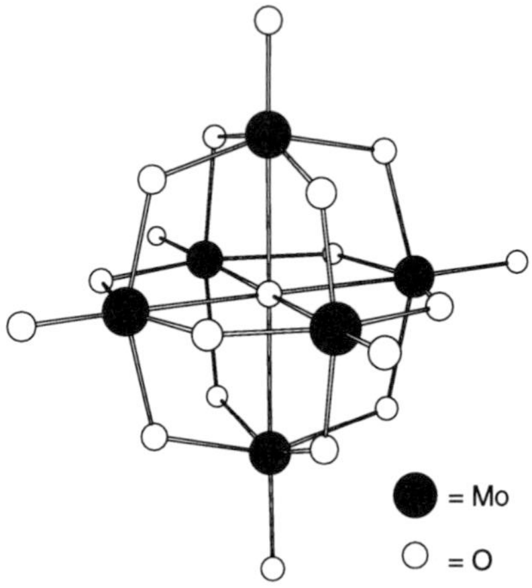

● = Mo	○ = O

Fig. 1.5 The structure of the $[Mo_6O_{19}]^{2-}$ dianion.

Iron-sulfur cubanes such as $[Fe_4S_4(SEt)_4]^{2-}$ (Fig. 1.6) are model compounds for iron-sulfur proteins. In the central Fe_4S_4-core, each sulfur atom bridges three iron centres but the unit is often distorted in such a way that the iron atoms move together in a pair-wise manner. The Fe···Fe internuclear distances indicate that there are no metal–metal bonds, but magnetic data show that there is communication between the iron centres, and in iron sulfur proteins, this is fundamental to their ability to exhibit a range of redox potentials.

Fig. 1.6 The structure of $[Fe_4S_4(SEt)_4]^{2-}$; the H atoms are omitted.

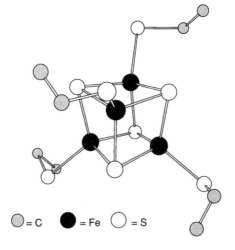

◐ = C	● = Fe	○ = S

In this book, we concentrate on clusters with π-acceptor ligands. Seeing a three-dimensional metal core in the molecule or ion immediately suggests that *cluster* is an appropriate descriptive term. However, there is a large group of compounds, the members of which possess a trinuclear core. In some species this is triangular (Fig. 1.7), but in others the M_3-core is open (Fig. 1.8). There is also an increasing number of compounds in which the metal core consists of triangular units which share common edges but in which the overall metal framework is approximately planar — the so-called 'rafts'. In this text, we include species with triangular metal-units, but exclude compounds with linear or bent M_3-units.

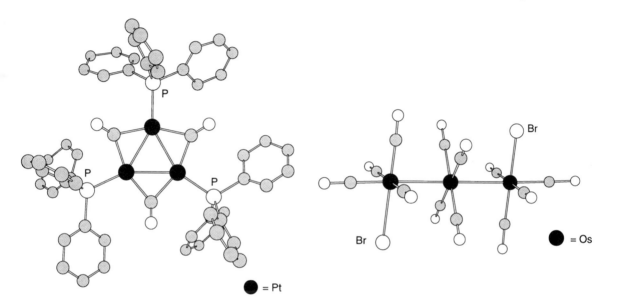

Fig. 1.7 The structure of $Pt_3(PPh_3)_3(\mu\text{-}CO)_3$.

Fig. 1.8 The structure of $Os_3(CO)_{12}Br_2$.

1.3 Some nomenclature

The modes by which ligands bind to metal centres are indicated in compound formulae by the use of the Greek letters μ and η, with, respectively, subscripted and superscripted numeric qualifiers. The symbol μ refers to bridging modes as follows:

These terms are discussed in more detail in the companion Primer *Organometallics 1: Complexes with Transition Metal–Carbon σ-Bonds* M. Bochmann, 1994.

μ or μ_2	Bridging 2 metal centres; edge bridging	e.g. μ-H
μ_3	Bridging 3 metal centres; capping a triangular face	e.g. μ_3-H

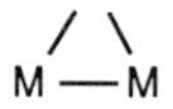

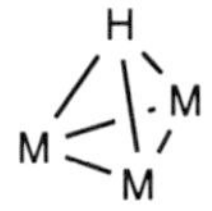

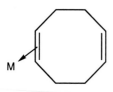

Fig. 1.9 Cycloocta-1,5-diene could coordinate to a metal centre in either an η^2- or η^4-mode. Note how the ring folds to accommodate the η^4-mode.

The symbol η refers to the *hapticity* of the ligand:

η^2 e.g. alkene

η^5 e.g. cyclopentadienyl

η^3 e.g. allyl

η^6 e.g. benzene

Figure 1.9 illustrates how the descriptor is used to distinguish between possible bonding modes of a versatile ligand.

1.5 Further reading

General texts

The Chemistry of Metal Cluster Complexes, (1990), eds. D.F. Shriver, H.D. Kaesz and R.D. Adams, VCH, New York. — This text includes a chapter by Mingos *et al.* on the structure and bonding of metal clusters.

Introduction to Cluster Chemistry, (1990), D.M.P. Mingos and D.J. Wales, Prentice Hall, New Jersey.

Reviews on specific topics

Alkyne-substituted transition metal clusters, (1985) P.R. Raithby and M.J. Rosales, *Advances in Inorganic and Radiochemistry*, vol. 29, 169-247.

The chemistry of carbidocarbonyl clusters, (1983) J.S. Bradley, *Advances in Organometallic Chemistry*, vol. 22, p. 1-58.

Organometallic metal clusters containing nitrosyl and nitrido ligands (1985) W.L. Gladfelter, *Advances in Organometallic Chemistry*, vol. 24, p. 41-86.

Transition metal–main group cluster compounds, (1992) C.E. Housecroft, in *Inorganometallic Chemistry*, ed. T.P. Fehlner, Plenum Press, New York, chapter 3, p. 73-178.

Coordination Chemistry Reviews, (1995) vol. 143, p. 1-678 — a special issue containing reviews, many with organometallic transition metal clusters.

High-nuclearity carbonyl clusters: Their synthesis and reactivity, (1986) M.D. Vargas and J.N. Nicholls, *Advances in Inorganic Chemistry and Radiochemistry*, vol. 30, p. 123–222.

Reflections on osmium and ruthenium carbonyl compounds, (1995) J. Lewis and P.R. Raithby, *Journal of Organometallic Chemistry*, vol. 500, 227-237.

Metal surfaces and catalysis

Transition metal clusters in homogeneous catalysis, (1993) G. Süss-Fink and G. Meister, *Advances in Organometallic Chemistry*, vol. 35, p. 41–134.

Proton induced reduction of CO to CH_4 in homonuclear and heteronuclear metal carbonyls, (1982) M.A. Drezdon, K.H. Whitmire, A.A. Bhattacharyya, W.-L. Hsu, C.C. Nagel, S.G. Shore and D.F. Shriver, *Journal of the American Chemical Society*, vol. 104, p. 5630-5633.

An organometallic guide to the chemistry of hydrocarbon moieties on transition metal surfaces, (1995) F. Zaera, *Chemical Reviews*, vol. 95, p. 2651-2693.

2 Binary metal carbonyls: synthesis, structures and localized bonding schemes

2.1 The 18-electron rule and dimer formation

A low oxidation state transition metal which is associated with π-acceptor ligands tends to obey the 18-electron rule (see Box 2.1) although the rule often breaks down for early and late metals in the d-block. Consider the manganese atom with a ground state electronic configuration of $[Ar]4s^2 3d^5$. Combined with five CO ligands, it forms a 17-electron radical $[Mn(CO)_5]^{\bullet}$. In order to complete the 18-electron configuration, the radical could accept an electron or dimerize (Fig. 2.1). Note the analogy with the methyl radical with which $[Mn(CO)_5]^{\bullet}$ is isolobal (Fig. 2.2).

Reduction of the dimer with sodium cleaves the Mn–Mn bond and leads to the formation of the $[Mn(CO)_5]^-$ anion (Eqn 2.1). Dihydrogen also reduces $Mn_2(CO)_{10}$ (Eqn 2.2).

The use of π-acceptor ligands to stabilize low oxidation state metal centres, application of the 18-electron rule, and isolobality in organometallic compounds are discussed in the companion Primer *Organometallics 1: Complexes with Transition Metal–Carbon σ-Bonds* M. Bochmann, 1994.

$$Mn_2(CO)_{10} + 2Na \longrightarrow 2Na^+[Mn(CO)_5]^- \qquad \text{Eqn 2.1}$$

$$Mn_2(CO)_{10} + H_2 \longrightarrow 2HMn(CO)_5 \qquad \text{Eqn 2.2}$$

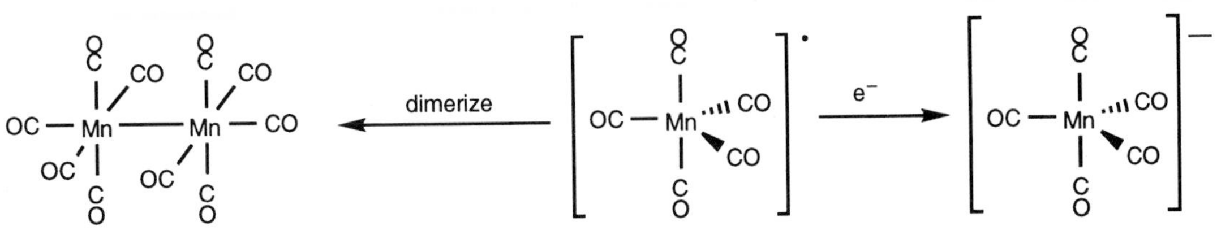

Fig. 2.1 The 17-electron $[Mn(CO)_5]^{\bullet}$ can be stabilized by the addition of an electron or by dimerization.

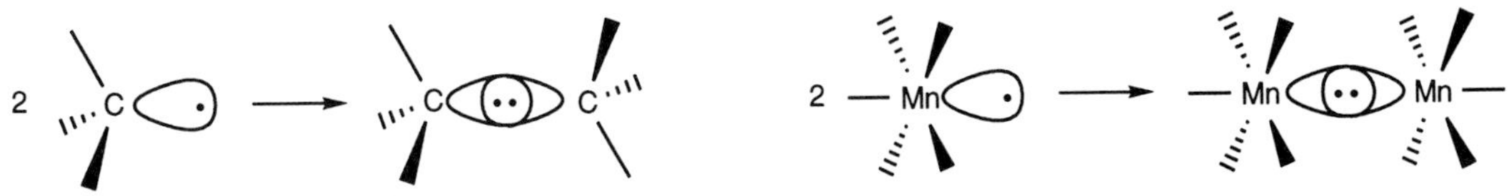

Fig. 2.2 $CH_3^{\bullet}$ and $[Mn(CO)_5]^{\bullet}$ are isolobal; each dimerizes by forming a 2-centre 2-electron C–C or Mn–Mn bond respectively.

Box 2.1   Rationalization of the 18-electron rule for an octahedral metal carbonyl complex

The HOMO and LUMO of a carbonyl ligand are :

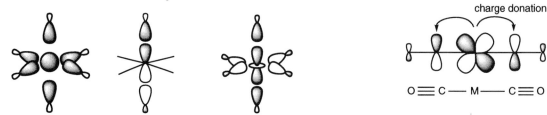

HOMO: σ-donor orbital   LUMO: π-acceptor orbital

In an octahedral complex such as $Cr(CO)_6$, the $4s$, $4p$ and the $3d_{z^2}$ and $3d_{x^2-y^2}$ orbitals of the chromium atom can overlap with the σ-donor orbitals of the octahedral set of six CO ligands. For example, the interaction of the $4s$, $4p_z$ and $3d_{z^2}$ AOs with the ligand orbitals are shown in the first three diagrams below:

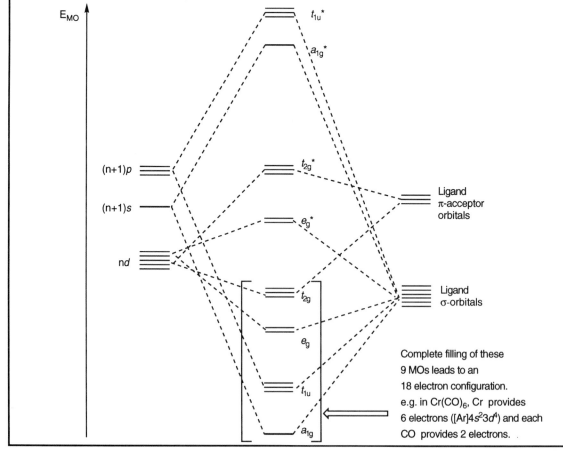

In $Cr(CO)_6$, the metal centre is formally in a zero oxidation state and the donation of electronic charge from the 6 CO ligands is countered (i.e. the electroneutrality principle) by *back donation* of electron density from the t_{2g} set of orbitals (the $3d_{xy}$, $3d_{xz}$ and $3d_{yz}$) to the lowest unoccupied MOs of the carbonyl ligands. The carbonyls thereby function as π-acceptor ligands as shown in the right-hand diagram above. The overall MO scheme for the formation of an octahedral species with donor and π-acceptor interactions is shown below. If there are sufficient valence electrons from the metal AND the six ligands to fill the 9 MOs indicated, all the bonding MOs are occupied and the 18-electron rule is obeyed.

Complete filling of these 9 MOs leads to an 18 electron configuration. e.g. in $Cr(CO)_6$, Cr provides 6 electrons ($[Ar]4s^2 3d^4$) and each CO provides 2 electrons.

The method of counting electrons in $Mn_2(CO)_{10}$ is as follows.
For each Mn centre:

$$Mn^0 \ [Ar]4s^23d^5 \ = 7 \text{ electrons}$$
$$5 \ CO = 10 \text{ electrons}$$
$$Mn–Mn \text{ bond} = 1 \text{ electron (2-centre 2-electron bond}$$
$$\text{shared between two Mn centres)}$$
$$Total = 18 \text{ electrons}$$

> **Important note**: The consideration of a dimer or cluster in terms of the combination of fragments is a useful way of discussing the bonding. However, in most cases, it is purely a *formalism* and does not represent a synthetic strategy.

In $Mn_2(CO)_{10}$, each manganese centre is octahedral, and rotation about the Mn–Mn single bond allows the molecule to adopt a sterically favourable staggered conformation in the solid state (Fig. 2.3). Similar structures are observed for both $Tc_2(CO)_{10}$ and $Re_2(CO)_{10}$; these may be prepared by reducing Tc_2O_7 or Re_2O_7 respectively (Eqn 2.3).

$$Re_2O_7 + 17CO \longrightarrow Re_2(CO)_{10} + 7CO_2 \qquad \textbf{Eqn 2.3}$$

Steric constraints are important in dimer formation. By using the 18-electron rule we could predict the formation of $V_2(CO)_{12}$; the ground state electronic configuration of V^0 is $[Ar]4s^23d^3$ and $V(CO)_6$ contains a 17-electron vanadium centre. However, dimerization by forming a V–V single bond, with all the CO ligands terminally bonded, would entail the vanadium atom becoming seven-coordinate and this is sterically unfavourable. In fact, monomeric $V(CO)_6$ may be isolated as air-sensitive, blue crystals which are paramagnetic and decompose at 335 K. The anion $[V(CO)_6]^-$ obeys the 18-electron rule.

Going from Mn to V in the first row of the *d*-block meant that the metal provided two *less* electrons for bonding and additional ligands were needed to fulfil the requirements of the 18-electron rule. If we move in the other direction and consider cobalt, $[Ar]4s^23d^7$, then only nine more electrons are needed for the formation of an 18-electron species. Thus we predict the formation of a dimeric species $Co_2(CO)_8$ and the structure shown in Fig. 2.4 is a reasonable proposal. In fact dicobalt octacarbonyl adopts this structure in solution but it, and at least one other directly Co–Co bonded isomer, are in equilibrium with the structure shown in Fig. 2.5 — it is this form that is observed in the solid state.

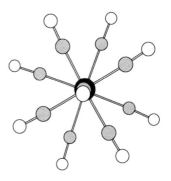

Fig. 2.3 The structure of $Mn_2(CO)_{10}$ in the solid state, viewed along the Mn–Mn axis to show the staggered conformation.

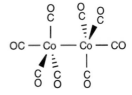

Fig. 2.4 An isomer of $Co_2(CO)_8$ which has a direct Co–Co bond and no bridging CO ligands.

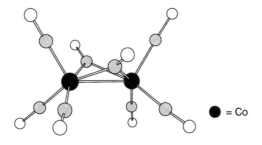

● = Co

Fig. 2.5 The solid state structure of $Co_2(CO)_8$, determined by X-ray crystallography.

Figure 2.5 illustrates an example of a dimer in which the metal–metal interaction is supported by bridging (μ) carbonyl ligands. In terms of formal electron counting, each μ-CO ligand provides one electron to each metal centre. The question arises — 'Is the dimer wholly supported by the μ-CO ligands or is there a Co–Co bond?'.

An electron count for each cobalt atom, *assuming no* Co–Co bond, is as follows:

$$
\begin{array}{rl}
\text{Co}^0 \ [\text{Ar}]3s^2 4d^7 & = 9 \text{ electrons} \\
3 \text{ CO (terminal)} & = 6 \text{ electrons} \\
2 \ \mu\text{-CO} & = 2 \text{ electrons} \\
\hline
\text{Total} & = 17 \text{ electrons}
\end{array}
$$

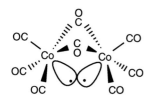

Fig. 2.6   The formation of a bent Co–Co bond in $Co_2(CO)_8$.

This result points towards the need for a Co–Co bond in order that each Co centre obeys the 18-electron rule. However, the orientations of the carbonyl ligands (as confirmed from X-ray crystallographic studies, Fig. 2.5) mean that the orbitals which are available for Co–Co bond formation do *not* point directly towards each other; compare Fig. 2.6 with Fig. 2.2. In terms of hybrid orbitals, each cobalt centre may be considered to be sp^3d^2 hybridized (consistent with an octahedral environment) — five hybrid orbitals are used for bonding to the CO ligands leaving one orbital for metal–metal bond formation. The result is a 'bent' Co–Co bond. The fact that $Co_2(CO)_8$ is diamagnetic supports the presence of a Co–Co bonding interaction, although it should be noted that the diamagnetic behaviour could also be rationalized in terms of electron spin coupling through the μ-CO ligands or antiferromagnetic coupling.

The ^{13}C NMR spectroscopic data show that $Co_2(CO)_8$ is fluxional even *in the solid state*. One mechanism is via terminal-bridge CO exchange and the activation energy for this process has been estimated to be ≈ 50 kJ mol^{-1}.

Dicobalt octacarbonyl is commercially available, but there are several methods of preparing it in the laboratory and one route is shown in Eqn 2.4.

$$
\text{Co(O}_2\text{CMe)}_2 \cdot 4\text{H}_2\text{O} \xrightarrow[\text{in acetic anhydride}]{\text{CO/H}_2 \ (4:1) \ 200 \ \text{atm, } 430\text{-}450 \ \text{K}} \text{Co}_2(\text{CO})_8 \qquad \textbf{Eqn 2.4}
$$

Rh$_4$(CO)$_{12}$ is described in Section 2.4.

The later members of group 9, rhodium and iridium, do not form stable dimers — Rh$_2$(CO)$_8$ is unstable with respect to disproportionation to Rh$_4$(CO)$_{12}$.

The dimerization of $[\text{Mn(CO)}_5]^{\bullet}$ or $[\text{Co(CO)}_4]^{\bullet}$ occurs because each is a 17-electron species. What happens if we consider the metal that lies between Mn and Co in the *d*-block? The ground state electronic configuration of Fe is $[\text{Ar}]4s^2 3d^6$ and the compounds Fe(CO)$_5$ and $[\text{Fe(CO)}_4]^{2-}$ are *monomeric*, 18-electron species. According to our definition in Chapter 1, the dinuclear carbonyl Fe$_2$(CO)$_9$ is not a dimer; the solid state structure (Fig. 2.7) confirms the presence of three terminal CO ligands per Fe atom with three bridging CO groups supporting the Fe–Fe interaction. The 18-electron rule suggests that an Fe–Fe bond is present, and this is consistent with the fact that the compound is diamagnetic. Nonetheless, the question of metal–metal bonding in $Fe_2(CO)_9$ has fuelled much work from a theoretical standpoint.

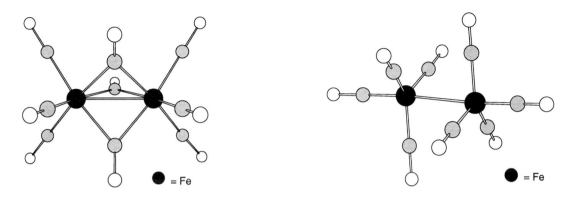

Fig. 2.7 The solid state structures of $Fe_2(CO)_9$ and $[Fe_2(CO)_8]^{2-}$. In $Fe_2(CO)_9$, if the Fe–Fe bond is ignored, each Fe centre is octahedrally coordinated. In $[Fe_2(CO)_8]^{2-}$, each Fe is in a trigonal bipyramidal environment; this description includes the Fe–Fe bond. The ligands in $[Fe_2(CO)_8]^{2-}$ adopt a staggered conformation.

The usual preparative route to $Fe_2(CO)_9$ involves the photolysis of $Fe(CO)_5$ (Eqn 2.5), and reduction of $Fe_2(CO)_9$ (Eqn 2.6) produces the dianion $[Fe_2(CO)_8]^{2-}$ which is isoelectronic with $Co_2(CO)_8$. An alternative route to $[Fe_2(CO)_8]^{2-}$ begins with $Fe(CO)_5$ (Eqn 2.7). In the solid state, $[Fe_2(CO)_8]^{2-}$ has the structure shown in Fig. 2.7, contrasting with that of its isoelectronic counterpart $Co_2(CO)_8$.

$$2Fe(CO)_5 \xrightarrow{h\nu} Fe_2(CO)_9 + CO \qquad\qquad \textbf{Eqn 2.5}$$

$$Fe_2(CO)_9 \xrightarrow[- CO]{KOH, [Et_4N]I} [Et_4N]_2[Fe_2(CO)_8] \qquad\qquad \textbf{Eqn 2.6}$$

$$Fe(CO)_5 + Na_2[Fe(CO)_4] \xrightarrow{THF} Na_2[Fe_2(CO)_8] + CO \qquad\qquad \textbf{Eqn 2.7}$$

Raman spectroscopy can be used as a means of probing for direct metal–metal bonding in dimetallic species. For first row transition metal carbonyl compounds, a direct M–M bond gives rise to an absorption in the region of 140-190 cm^{-1}, but this characteristically shifts above 200 cm^{-1} when the M–M interaction involves bridging CO ligands. The Raman spectrum for $[Et_4N]_2[Fe_2(CO)_8]$ has an absorption at 160 cm^{-1} (consistent with the Fe–Fe bond shown in Fig. 2.7) but this frequency is cation dependent. In DMF solution, $[MeH_2N(CH_2)_3NH_2Me][Fe_2(CO)_8]$ shows Raman absorptions at both 222 and 167 cm^{-1} demonstrating the presence of two isomers of $[Fe_2(CO)_8]^{2-}$, one bridged and one non-bridged.

2.2 Bond dissociation enthalpies

Before progressing to larger systems, we should consider the strength of M–M bonds. Despite the large number of organometallic compounds containing metal–metal bonds, there are few relevant and *reliable* M–M bond dissocation enthalpy data. In the bulk state, the strengths of metal–metal bonds can be estimated from standard enthalpies of atomization, but the structures of the transitions metals vary and partitioning of a molar enthalpy of atomization into bond dissociation enthalpies is structure dependent. The variation in $\Delta H^\circ_{\text{atomization}}$ (298 K) for the transition metals is shown in Fig. 2.8 and can be used to give some idea of the variation in metal–metal bond dissociation enthalpies. However, a word of caution — these values refer to the zero valent state. In many carbonyl and related compounds, the metals are also zero valent, but compounds with π-donor ligands (e.g. halide, sulfide, oxide) contain metals in higher oxidation states. Since it is the earlier metals (with fewer valence electrons) that tend to form complexes with π-donor ligands and the later metals (with more valence electrons) that combine with π-acceptor ligands, we concentrate here on the right-hand side of Fig. 2.8.

Two general trends should be noted. Firstly, the enthalpy values for the third row metals are higher than those of the first and second rows, and in general, the values for the second row metals are greater than those of the first row elements. This is consistent with the greater orbital overlap that exists between AOs of higher principal quantum numbers $5d$-$5d > 4d$-$4d > 3d$-$3d$). Secondly, for the second and third rows, the highest values are found for the metals in the middle groups of the d-block. This appears to be reflected in the fact that the most well established *organometallic* cluster chemistry and the formation of high nuclearity species involve Ru and Os (group 8), and to a lesser extent Rh and Ir (group 9).

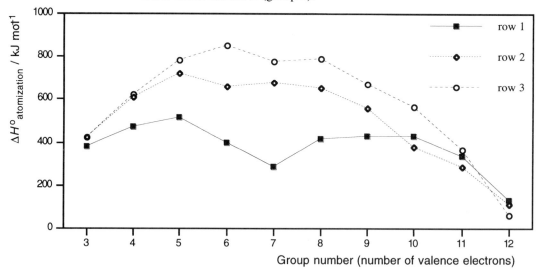

Fig. 2.8   The periodic trends in molar $\Delta H^\circ_{\text{atomization}}$ (298 K) for the d-block metals arranged by row.

2.3 Trinuclear compounds

2.3.1 Neutral compounds

The group 8 metals form a family of carbonyls of the type $M_3(CO)_{12}$. Preparative methods for the dark green compound $Fe_3(CO)_{12}$ vary; one route is the oxidation of $[HFe(CO)_4]^-$ by MnO_2. Of the routes available to $Ru_3(CO)_{12}$ (an orange crystalline solid), the carbonylation of $RuCl_3$ is an efficient method (Eqn 2.8) although the product can be contaminated with $H_4Ru_4(CO)_{12}$. Yellow crystals of $Os_3(CO)_{12}$ are formed by the reaction in an autoclave of OsO_4 with CO (Eqn 2.9); other osmium clusters are formed as by-products.

$H_4Ru_4(CO)_{12}$ is described in Section 4.2.

$$RuCl_3 \cdot xH_2O + CO \xrightarrow{\text{MeOH, 400K, 50atm}} Ru_3(CO)_{12} \qquad \textbf{Eqn 2.8}$$

$$OsO_4 + CO \xrightarrow{\text{MeOH, 400 K, } \leq 200\text{atm}} Os_3(CO)_{12} \qquad \textbf{Eqn 2.9}$$

The structures of $Ru_3(CO)_{12}$ and $Os_3(CO)_{12}$ are shown in Fig. 2.9. In each compound, all the CO ligands are terminally bonded, and Fig. 2.9 illustrates that the formation of single metal–metal bonds can readily be rationalized in terms of the 18-electron rule; the pictorial representation of the $M(CO)_4$ fragment assumes an sp^3d^2 hybridization scheme, consistent with the approximately octahedral environment in which each metal atom is observed in $Ru_3(CO)_{12}$ and $Os_3(CO)_{12}$.

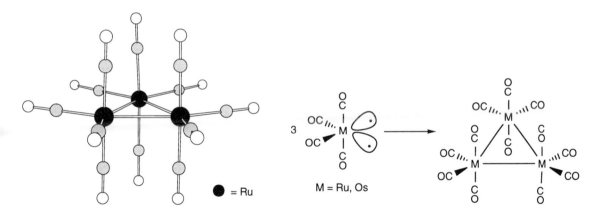

Fig. 2.9   The solid state structure of $Ru_3(CO)_{12}$; $Os_3(CO)_{12}$ is isostructural with $Ru_3(CO)_{12}$. Each $M(CO)_4$ unit is a 16 electron centre and forms two metal–metal single bonds, thereby obeying the 18-electron rule.

Triangular M_3-units are commonly observed, not only in trinuclear compounds, but also within higher nuclearity clusters such as those with tetrahedral and octahedral geometries. Any ring-strain effects that might be associated with the 60° angles of an equilateral M_3-triangle are reduced because of the involvement of the metal d-orbitals. These orbitals are relatively diffuse and interaction between them is not restricted to the metal–metal vectors indicated in, for example, Fig. 2.9.

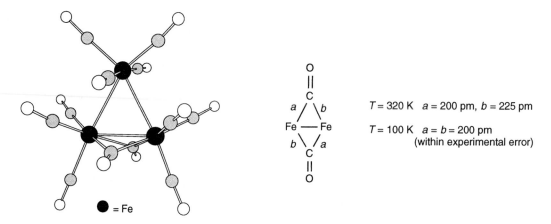

Fig. 2.10   The solid state structure of $Fe_3(CO)_{12}$; the bridging unit is asymmetrical at 320 K but virtually symmetrical at 100 K. An alternative way of writing the formula is $Fe_3(CO)_{10}(\mu\text{-}CO)_2$ — this gives more structural information.

Remember that each μ-CO contributes one electron to each metal centre.

There is an interesting contrast between the structure of $Fe_3(CO)_{12}$ and those of the ruthenium and osmium analogues. In the solid state, $Fe_3(CO)_{12}$ possesses the structure shown in Fig. 2.10 — two CO ligands bridge one Fe–Fe edge of the metal triangle, and crystallographic studies at several temperatures have revealed that these bridges are asymmetrical at 320 K but become symmetrical by 100 K. Each Fe centre obeys the 18-electron rule if three 2-centre 2-electron Fe–Fe bonds are invoked.

In solution, $Fe_3(CO)_{12}$ shows only one ^{13}C NMR spectroscopic resonance even at 123 K, indicating that the molecule is stereochemically non-rigid. Several mechanisms have been proposed to account for these data. Bridge-terminal CO exchange permits the migration of the ligands around the Fe_3-core. Alternatively, the molecule can be considered in terms of an Fe_3-unit within an icosahedral cage of CO ligands. There are three possible orientations that the Fe_3-triangle could adopt within this cage and Fig. 2.11 illustrates the situation for the crystallographically characterized structure. If the Fe_3-unit tilts within the $(CO)_{12}$-cage, a mechanism can be developed that renders the CO ligands equivalent. Such motions are low energy processes and the dynamic behaviour of $Fe_3(CO)_{12}$ *in the solid state* may also be interpreted in terms of this model.

Fig. 2.11   A space-filling diagram of the solid state structure of $Fe_3(CO)_{12}$ shows that the CO ligands define an approximately icosahedral shell around the Fe_3-core.

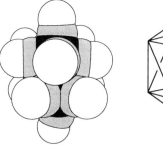

Icosahedron

2.3.2 Anions

When $Co_2(CO)_8$ reacts with an alkali metal salt of $[Co(CO)_4]^-$, the product is a salt of $[Co_3(CO)_{10}]^-$ (Eqn 2.10). Removal of CO from the equilibrium mixture results in further reaction and the formation of $Co_4(CO)_{12}$ (see Section 2.4.1).

$$Co_2(CO)_8 + K[Co(CO)_4] \rightleftharpoons K[Co_3(CO)_{10}] + 2CO \qquad \textbf{Eqn 2.10}$$

The structure of the $[Co_3(CO)_{10}]^-$ anion is shown in Fig. 2.12, and the the presence of the three different types of CO ligand in $[Co_3(CO)_{10}]^-$ is readily deduced from the infrared spectrum which exhibits absorptions at 2071, 2006, 1999, 1975, 1865 and 1584 cm^{-1} (Fig. 12.13).

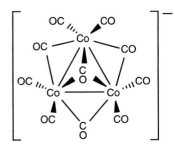

Fig. 2.12 The solid state structure of $[Co_3(CO)_{10}]^-$, determined for the lithium salt.

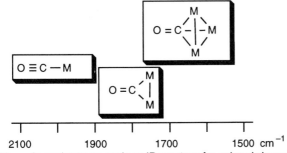

Fig. 2.13 Approximate ranges in an IR spectrum for carbonyl absorptions in metal carbonyl compounds.

Further details of the use of IR spectroscopy are given in the companion Primer *Organometallics 1: Complexes with Transition Metal–Carbon σ-Bonds* M. Bochmann, 1994.

The electron count at each metal centre in $[Co_3(CO)_{10}]^-$ can be determined as follows:

$$
\begin{array}{rl}
Co^0 \ [Ar]4s^23d^7 & = 9 \text{ electrons} \\
2 \text{ CO (terminal)} & = 4 \text{ electrons} \\
2 \ \mu\text{-CO} & = 2 \text{ electrons (one from each CO)} \\
\mu_3\text{-CO} & = {}^2/_3 \text{ electron} \\
1- \text{ charge (shared over 3} & \\
\text{Co centres)} & = {}^1/_3 \text{ electron} \\
2 \text{ Co–Co bonds} & = 2 \text{ electrons} \\
\text{Total} & = 18 \text{ electrons}
\end{array}
$$

Such a counting scheme is rather clumsy, and as we see later, it is often easier to consider the metal framework as a whole rather than localize electrons at each metal centre.

The reaction between $Fe_3(CO)_{12}$ and hydroxide ion leads to the formation of $[Fe_3(CO)_{11}]^{2-}$ (Eqn 2.11); the product can be precipitated from solution with a large cation such as $[N(PPh_3)_2]^+$.

$$Fe_3(CO)_{12} + 4KOH \longrightarrow K_2[Fe_3(CO)_{11}] + K_2CO_3 + 2H_2O \qquad \textbf{Eqn 2.11}$$

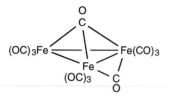

Fig. 2.14 The solid state structure of $[Fe_3(CO)_{11}]^{2-}$.

The structure of $[Fe_3(CO)_{11}]^{2-}$ is related to that of $Fe_3(CO)_{12}$, but there is one less terminal CO ligand, and one bridging CO adopts a μ_3-bonding mode (Fig. 2.14). The solid state and solution IR spectra of $[Fe_3(CO)_{11}]^{2-}$ confirm the presence of the μ_3-CO ligand with an absorption at 1667 cm^{-1}. On the ^{13}C NMR spectroscopic timescale, all the CO ligands are involved in an exchange process which persists at 173 K, and only one ^{13}C NMR resonance is observed (δ +231).

2.4   Tetranuclear binary metal carbonyl clusters

2.4.1   Tetrahedral clusters

We have already seen that the $Co(CO)_4$ fragment is a 17-electron unit and readily dimerizes. The removal of one CO ligand generates a 15-electron metal centre which has the potential for forming three metal–metal single bonds, and aggregation of these fragments gives the tetrameric cluster $Co_4(CO)_{12}$. Similarly, $Rh_4(CO)_{12}$ and $Ir_4(CO)_{12}$ can be constructed 'on paper' from the combination of four $M(CO)_3$ units (Fig. 2.15). Each cluster possesses a *tetrahedral* M_4-framework.

In practice, $Co_4(CO)_{12}$ is formed by the thermal decomposition of $Co_2(CO)_8$, but for the heaver group 9 metals, the reactions shown in Eqns 2.12 and 2.13 may be used.

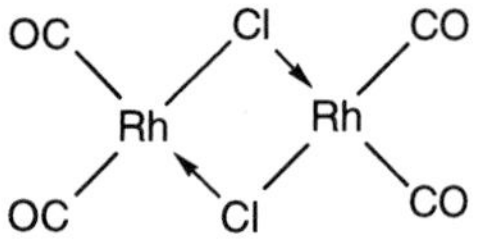

$$Rh_2(\mu\text{-Cl})_2(CO)_4 \xrightarrow[\text{NaHCO}_3]{\text{hexane, CO 1 atm, 298 K}} Rh_4(CO)_{12} \qquad \textbf{Eqn 2.12}$$

$$Na_3[IrCl_6] \xrightarrow[\text{2. base}]{\text{1. NaI in MeOH under reflux, CO 1 atm}} Ir_4(CO)_{12} \qquad \textbf{Eqn 2.13}$$

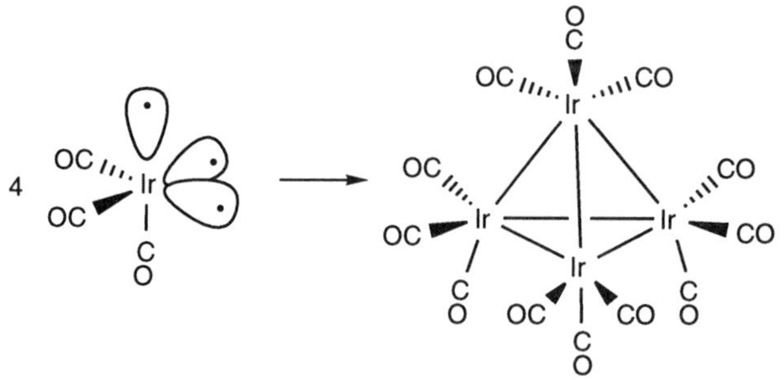

$Rh_2(\mu\text{-Cl})_2(CO)_4$

Fig. 2.15 The formation of a tetrahedral cluster from four $Ir(CO)_3$ units; each unit is approximately octahedral and an sp^3d^2 hybridization scheme is appropriate. Each Ir centre obeys the 18-electron rule.

The crystallographically determined structure of $Ir_4(CO)_{12}$ confirms the presence of 12 terminal carbonyl ligands but in $Co_4(CO)_{12}$ and $Rh_4(CO)_{12}$ (which are isostructural, Fig. 2.16) the ligand distribution is different: 9 terminal CO and 3 μ-CO groups. Figure 2.17 illustrates the polyhedral arrangements of the carbonyl ligands in $Co_4(CO)_{12}$, $Rh_4(CO)_{12}$ and $Ir_4(CO)_{12}$. In the same way that we viewed the structure of $Fe_3(CO)_{12}$ in terms of an Fe_3-triangle within an icosahedron of CO ligands, the structures of group 9 tetrametal carbonyl clusters can be considered in terms of a metal tetrahedron within either an icosahedron or cubeoctahedron of carbonyls.

In solution, $Co_4(CO)_{12}$ and $Rh_4(CO)_{12}$ are fluxional; at 336 K the ^{13}C NMR spectrum of $Rh_4(CO)_{12}$ shows a binomial quintet indicating that, on the NMR spectroscopic timescale, all the CO ligands are equivalent and each ^{13}C nucleus couples to all four ^{103}Rh nuclei. The fluxionality is frozen out at ≈ 200 K, and the ^{13}C NMR spectrum is in accordance with the solid state structure. The solution spectra of $Rh_4(CO)_{12}$ can be explained in terms of terminal-bridge carbonyl exchange or by the motion of the Rh_4-tetrahedron inside a polyhedral shell of 12 CO ligands.

^{103}Rh is an NMR active nucleus 100% $I = \frac{1}{2}$.

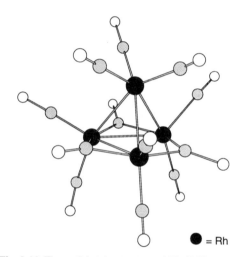

Icosahedron   Cubeoctahedron

Fig. 2.16 The solid state structure of $Rh_4(CO)_{12}$; $Co_4(CO)_{12}$ adopts the same structure, but in $Ir_4(CO)_{12}$, all the CO ligands are terminally bonded.

● = Rh

Fig. 2.17 In $Co_4(CO)_{12}$ and $Rh_4(CO)_{12}$, the CO ligands define an approximate icosahedron but in $Ir_4(CO)_{12}$ they define a cubeoctahedron.

When a solution of $Fe(CO)_5$ in pyridine is heated to reflux, the salt $[Fe(py)_6][Fe_4(CO)_{13}]$ is formed; metathesis with $[(Ph_3P)_2N]Cl$ gives the more stable derivative $[(Ph_3P)_2N]_2[Fe_4(CO)_{13}]$. The $[Fe_4(CO)_{13}]^{2-}$ dianion possesses a tetrahedral skeleton of metal atoms. One CO ligand adopts a μ_3-bonding mode (IR $v(CO)$ 1661 cm^{-1}) and Fig. 12.18 shows a view of the molecule looking along the O–C vector of the μ_3-CO group. This orientation of the $[Fe_4(CO)_{13}]^{2-}$ ion reveals that there are three *semi-bridging* CO ligands, a mode which is midway between terminal and fully bridging. The Fe–C bond lengths of 180 and 220 pm for each of these ligands are consistent with the carbonyl group being more weakly associated with one Fe centre than the other. In solution, the ^{13}C NMR spectrum of $[Fe_4(CO)_{13}]^{2-}$ (even at 173 K) shows only one carbonyl environment, indicating the existence of a fluxonal process which renders all the carbonyl ligands equivalent on the NMR spectroscopic timescale.

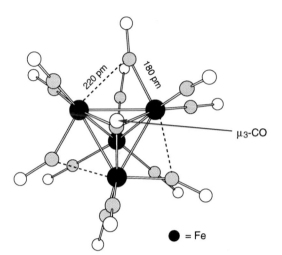

Fig. 2.18   The $[Fe_4(CO)_{13}]^{2-}$ anion from the solid state structure of $[Fe(py)_6][Fe_4(CO)_{13}]$.

It has been argued that steric crowding of CO ligands may be the reason why '$Fe_4(CO)_{14}$' is unknown. The tetrahedral cluster $Os_4(CO)_{14}$ has been fully characterized — the Os_4-core is larger than its iron analogue would be and steric interactions between the ligands are therefore less in $Os_4(CO)_{14}$ than in a structurally similar '$Fe_4(CO)_{14}$'.

2.4.2   Square frameworks

The construction *on paper* of tetrahedral $M_4(CO)_{12}$ clusters in which M is a group 9 metal uses a building-block approach with $M(CO)_3$ fragments, each of which can form three M–M bonds. Moving to group 8, we have already seen that $M(CO)_4$ units are capable of forming two M–M bonds and trimeric compounds are observed. But surely, larger cyclic structures are also possible? In practice this is hardly ever observed — one example is $Os_4(CO)_{16}$ (Fig. 2.19) in which the four Os atoms adopt a puckered-square geometry and all the CO ligands are terminally bonded. Each Os centre obeys the 18-electron rule. The Os–Os bonds are particularly long ($\approx$300 pm) indicating that the metal–metal bonding is weak. This is supported by the observation that, on standing in hexane, $Os_4(CO)_{16}$ decomposes to $Os_3(CO)_{12}$. The preparative route to $Os_4(CO)_{16}$ is via $Os_4(CO)_{15}$ (Eqn 2.14) which has a raft structure (see below).

$$Os_3(CO)_9(COEt)_2 + Os(CO)_5 \xrightarrow[260\ K]{CH_2Cl_2,\ hexane} Os_4(CO)_{15} \qquad \textbf{Eqn 2.14}$$

$$Os_4(CO)_{15} \xrightarrow[273\ K]{CO\ (1\ atm)} Os_4(CO)_{16}$$

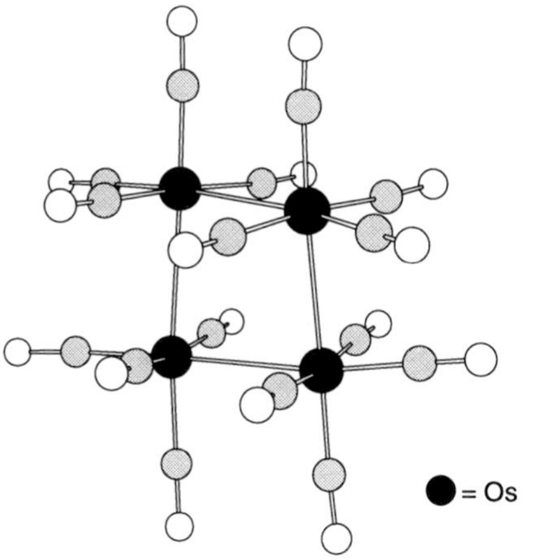

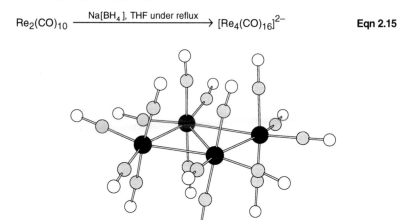

Fig. 2.19 The solid state structure of Os$_4$(CO)$_{16}$.

2.4.3 Planar rafts

Whilst the tetrahedron is by far the most common geometry for tetranuclear metal clusters, several 'planar raft' species are now known. A raft is composed of a series of edge-sharing triangles and may be larger than a tetranuclear framework (e.g. Os$_5$(CO)$_{18}$, Fig. 2.22). Two examples of four-atom rafts are Os$_4$(CO)$_{15}$ (Eqn 2.14) and the rhenium cluster [Re$_4$(CO)$_{16}$]$^{2-}$ (Fig. 2.20). The Re$_4$-framework consists of two edge-sharing triangles and each rhenium centre is bonded to four terminal CO ligands. This anion can be synthesized by reducing Re$_2$(CO)$_{10}$ (Eqn 2.15).

$$Re_2(CO)_{10} \xrightarrow{\text{Na[BH}_4\text{], THF under reflux}} [Re_4(CO)_{16}]^{2-} \qquad \textbf{Eqn 2.15}$$

Fig. 2.20 The [Re$_4$(CO)$_{16}$]$^{2-}$ dianion from the solid state structure of [^{n}Bu$_4$N]$_2$[Re$_4$(CO)$_{16}$].

2.5 Selected clusters with five or more metal atoms

The wide range of fully characterized transition metal carbonyl clusters containing five or more metal atoms precludes a comprehensive look at this group of compounds. In this section, examples have been chosen to illustrate some of the geometries observed and will allow us to see why the bonding in these species cannot reasonably be approached by allocating localized metal–metal bonds. It is left to the reader to apply the 18-electron rule to each metal

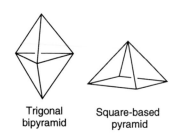

Trigonal bipyramid   **Square-based pyramid**

Fig. 2.21 The trigonal bipyramidal and square-based pyramidal geometries which may be adopted by M_5-cluster cores.

centre in each of the clusters described in this section — bonding schemes which circumvent the problems arising in this exercise are considered in Chapter 3.

2.5.1 Pentametal frameworks

Two common core geometries for five metal atoms are the trigonal bipyramid and the square-based pyramid (Fig. 2.21). However, such pentametal species are not well represented within the *binary* metal carbonyl clusters.

The osmium cluster $Os_5(CO)_{16}$ is formed as a minor product from the pyrolysis of $Os_3(CO)_{12}$ (Eqn 2.16) and possesses a trigonal bipyramidal framework. All the CO ligands are terminally bonded; one equatorial Os atom has four carbonyl groups attached.

$$Os_3(CO)_{12} \xrightarrow{\text{483 K}} Os_5(CO)_{16} + Os_6(CO)_{18} + Os_7(CO)_{21} + Os_8(CO)_{23}$$

major product   **Eqn 2.16**

The anions $[Rh_5(CO)_{15}]^-$ and $[Ni_5(CO)_{12}]^{2-}$ (Eqns 2.17 and 2.18) also possess trigonal bipyramidal metal cores. Crystallographic data for $[Rh_5(CO)_{15}]^-$ confirm that five CO ligands bridge Rh–Rh edges, but in solution (≈ 298 K), NMR spectroscopic data indicate that the anion is stereochemically non-rigid. In the solid state, the apical Ni atoms in $[Ni_5(CO)_{12}]^{2-}$ carry three terminal CO ligands, and each equatorial atom, one. Each Ni–Ni edge in the equatorial plane is supported by a μ-CO ligand.

$$Rh_4(CO)_{12} + [Rh(CO)_4]^- \xrightarrow{\text{CO, THF}} [Rh_5(CO)_{15}]^-$$ **Eqn 2.17**

$$Ni(CO)_4 \xrightarrow[\text{e.g. alkali metal}]{\text{reduction}} [Ni_5(CO)_{12}]^{2-} + [Ni_6(CO)_{12}]^{2-}$$ **Eqn 2.18**

In Eqn 2.19, the amine oxide oxidizes the terminal CO to CO_2 and releases the ligand from the metal:

$$CO + Me_3\overset{+}{N}-\overset{-}{O} \rightarrow CO_2 + Me_3N$$

Acetonitrile occupies the vacant coordination site on the metal centre:

$$M-CO \xrightarrow{-\text{CO}} [M]$$

$$[M] \xrightarrow{\text{MeCN}} M-NCMe$$

MeCN is a more labile ligand than CO.

Replacing one or more carbonyl groups by more labile ligands provides compounds which may be valuable precursors in cluster synthesis. One example is in the use of $Os_3(CO)_{10}(NCMe)_2$ (Eqn 2.19) which reacts with $H_2Os_3(CO)_{10}$ to form the pentaosmium cluster $Os_5(CO)_{18}$ (Fig. 2.22).

$$Os_3(CO)_{12} \xrightarrow{\text{Me}_3\text{NO, MeCN}} Os_3(CO)_{10}(NCMe)_2$$ **Eqn 2.19**

The Os_5-cluster core is a planar raft consisting of three edge-sharing triangles. All except one CO ligand are terminally bonded. The reaction of $Os_6(CO)_{18}$ (see section 2.5.2) with CO (90 atm) at elevated temperature leads to the formation of $Os_5(CO)_{19}$ which has an unusual 'bow-tie' metal framework (Fig. 2.23).

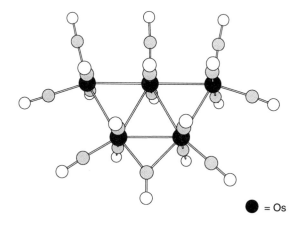

Fig. 2.22 The solid state structure of $Os_5(CO)_{18}$. The Os_5-core is planar.

● = Os

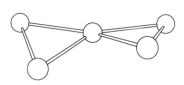

Fig. 2.23 The 'bow-tie' Os_5-core of $Os_5(CO)_{19}$.

2.5.2 Hexametal frameworks

Two well established hexametal frameworks are shown in Fig. 2.24, but it is the octahedral cage that dominates amongst clusters with six metal atoms. One standard view is shown in Fig. 2.24, but it is useful to remember that the octahedron can be considered in terms of two staggered triangular units (Fig. 2.25). Another point to note is that *all the vertices in a regular octahedron are equivalent.*

Reduction of $M_3(CO)_{12}$ (M = Ru or Os) gives the dianions $[M_6(CO)_{18}]^{2-}$ (Eqns 2.20 and 2.21) in which the M_6-framework is octahedral. Other examples of octahedral clusters are $Co_6(CO)_{16}$, $Rh_6(CO)_{16}$, $Ir_6(CO)_{16}$, $[Rh_6(CO)_{15}]^{2-}$, $[Rh_6(CO)_{14}]^{4-}$, $[Ir_6(CO)_{15}]^{2-}$ and $[Ni_6(CO)_{12}]^{2-}$ (Eqns 2.18 and 2.22).

$$Ru_3(CO)_{12} \xrightarrow{\text{Na, THF, } \Delta} [Ru_6(CO)_{18}]^{2-} \qquad \textbf{Eqn 2.20}$$

$$Os_3(CO)_{12} \xrightarrow{\text{Na, diglyme, } \Delta} [Os_6(CO)_{18}]^{2-} \qquad \textbf{Eqn 2.21}$$

$$3[Ni_5(CO)_{12}]^{2-} + 2H_2O \rightarrow 2[Ni_6(CO)_{12}]^{2-} + H_2 + 2OH^- + 3Ni(CO)_4 \qquad \textbf{Eqn 2.22}$$

The neutral cluster $Os_6(CO)_{18}$ (formed in reaction 2.16) does not have an octahedral geometry, but instead possesses the *condensed polyhedral* structure shown in Fig. 2.26. Five osmium atoms form a trigonal bipyramid and the sixth metal atom caps one triangular face. The Os_6-core can be considered in terms of a trigonal bipyramid and a tetrahedron which share one triangular face. An alternative description of this geometry is a *bicapped tetrahedron*. Condensed polyhedral structures become more prevalent as the nuclearity of the metal cluster increases ($>M_6$).

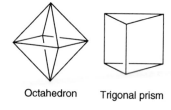

Octahedron   Trigonal prism

Fig. 2.24 The octahedral and trigonal prismatic geometries which may be adopted by M_6-cluster cores.

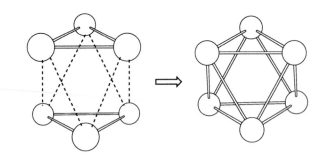

Fig. 2.25 The octahedron can be considered in terms of two staggered triangular units.

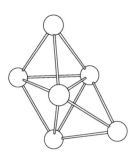

Fig. 2.26 The Os_6-core of $Os_6(CO)_{18}$.

In Fig. 2.25, if the two triangles were eclipsed rather than staggered, the resultant geometry would be a trigonal prism (Fig. 2.24). This structure is observed for $[Pt_6(CO)_{12}]^{2-}$ which belongs to a family of platinum carbonyls in which triangular Pt_3-units are arranged in stacks (see section 2.5.4). Amongst binary metal carbonyls, the trigonal prism is rare, and is better represented when the metal cavity is occupied by an interstitial atom, such as B, C, N, or P.

Clusters with interstitial atoms are described in Chapter 7.

2.5.3 Hepta- and octametal frameworks

Once above six metal atoms, the metal cluster geometries show a tendency to be composed of condensed tetrahedral and octahedral sub-clusters. As the metal cages become more and more complex, we shall focus attention on the metal core structures, and pay less attention to the disposition of the CO ligands. Where both terminal and bridging CO environments are present in the solid state, fluxional behaviour involving exchange of these sites tends to be observed in solution.

In a *localized exchange* process, ligands exchange positions at one metal centre but do not migrate from one metal to another.

Equation 2.16 showed the formation of $Os_7(CO)_{21}$ and $Os_8(CO)_{23}$ from $Os_3(CO)_{12}$. Crystallographic data for $Os_7(CO)_{21}$ confirm that the metal atoms define a *monocapped octahedron* (Fig. 2.27) — i.e. a tetrahedron and an octahedron which share one triangular face. All the CO ligands are terminally bonded, three per osmium atom. In solution from 273 to 363 K, the ^{13}C NMR spectrum of $Os_7(CO)_{21}$ exhibits three signals in a ratio 3:3:1, consistent with the presence of terminal CO ligands bound to three different metal sites. At each site, there is *localized exchange* of the ligands. The anion $[Rh_7(CO)_{16}]^{3-}$ also possesses a capped octahedral structure.

The dianion $[Os_8(CO)_{22}]^{2-}$ is formed from $Os_8(CO)_{23}$ by treatment with iodide ion, and possesses a *bicapped octahedral* structure. There are three possible isomers for the Os_8-core, labelled **a**, **b** and **c** in Fig. 2.28. The crystallographically confirmed structure of $[Os_8(CO)_{22}]^{2-}$ illustrates that geometry **b** is favoured. An example of structure **a** is found in the anion $[Re_6(CO)_{24}C]^{2-}$ which contains an interstitial carbon atom. Isomer **c** would appear to be sterically unfavourable, although it is found for the Ru_6Au_2-core of the heterometallic cluster $HRu_6(CO)_{16}B(AuPPh_3)_2$ in which the gold atoms adopt the capping sites.

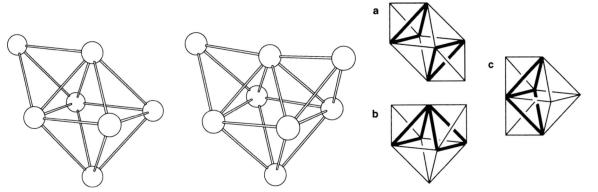

Fig. 2.27 The Os$_7$-core in Os$_7$(CO)$_{21}$.

Fig. 2.28 The crystallographically determined Os$_8$-core in [Os$_8$(CO)$_{22}$]$^{2-}$, and the three possible geometrical isomers of the bicapped octahedron.

An octametal framework can also be achieved by the coupling of two tetrahedral units by localized M-M bond formation, and this is observed in the anion [Ir$_8$(CO)$_{22}$]$^{2-}$ (Fig. 2.29). Formally, this cluster can be derived from Ir$_4$(CO)$_{12}$; the removal of one CO ligand to give {Ir$_4$(CO)$_{11}$} leaves one iridium atom coordinatively unsaturated (a 16-electron centre). Dimerization accompanied by the addition of two electrons regains the 18-electron count at each metal centre. In practice, the anion is synthesised according to Eqn 2.23.

$$Ir_4(CO)_{12} \xrightarrow[\text{2. [Et}_4\text{N]Cl}]{\text{1. Na, CO, THF}} [Et_4N]_2[Ir_8(CO)_{22}] \qquad \text{Eqn 2.23}$$

2.5.4 Clusters with nine or more metal atoms

The cluster anions [Rh$_9$(CO)$_{19}$]$^{3-}$, [Ni$_9$(CO)$_{18}$]$^{2-}$, and [Pt$_9$(CO)$_{18}$]$^{2-}$ shown in Fig. 2.30 form an interesting series; the nine metal atoms in each anion adopt a condensed polyhedral structure. In [Rh$_9$(CO)$_{19}$]$^{3-}$, the cage is composed of two face-sharing octahedra. The structure of [Ni$_9$(CO)$_{18}$]$^{2-}$ consists of a trigonal prism fused via a triangular face to an octahedron, whilst in [Pt$_9$(CO)$_{18}$]$^{2-}$, the metal framework comprises two face-sharing trigonal prisms. The M$_9$-cores of all three anions can be considered to be constructed from three stacked M$_3$-units, with the relative orientations of the triangles varying across the series.

The dianion [Pt$_9$(CO)$_{18}$]$^{2-}$ belongs to a series of platinum clusters of general formula [{Pt$_3$(CO)$_6$}$_n$]$^{2-}$ (n = 2 to 6). These anions are formed by the reductive carbonylation of Na$_2$[PtCl$_6$].6H$_2$O, but selective synthesis of a particular anion is difficult. Oxidation of a mixture of the anions with Na$_2$[PtCl$_6$] gives a route to [Pt$_{18}$(CO)$_{36}$]$^{2-}$, whilst reduction of the mixture with excess lithium yields [Pt$_6$(CO)$_{12}$]$^{2-}$. The structures of these clusters consist of stacks of {Pt$_3$(CO)$_6$}-units, each with three terminal and three μ-CO ligands. The Pt$_3$-triangles are stacked in an almost eclipsed configuration giving approximately trigonal prismatic units but the stack is twisted such that a helical assembly is observed (Figs. 2.30c and 2.31).

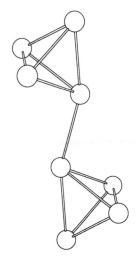

Fig. 2.29 The Ir$_8$-core of the [Ir$_8$(CO)$_{22}$]$^{2-}$ dianion.

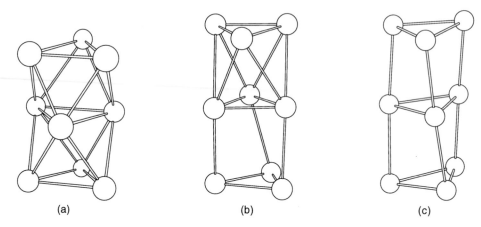

(a)   (b)   (c)

Fig. 2.30 The crystallographically confirmed structures of the M_9-frameworks of (a) $[Rh_9(CO)_{19}]^{3-}$, (b) $[Ni_9(CO)_{18}]^{2-}$, and (c) $[Pt_9(CO)_{18}]^{2-}$.

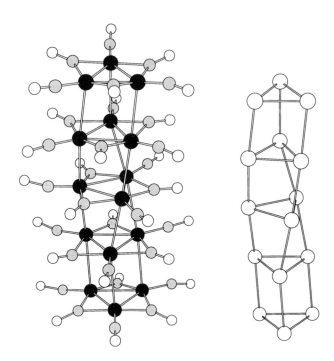

Fig. 2.31 The structure of $[Pt_{15}(CO)_{30}]^{2-}$ showing the stacked $Pt_3(CO)_6$-units; each unit has three terminal and three μ-CO ligands. The Pt_{15}-core possesses a helical twist.

The anion $[Rh_9(CO)_{19}]^{3-}$ is prepared by the cage-condensation of $[Rh_5(CO)_{15}]^-$ and $[Rh_4(CO)_{11}]^{2-}$ with accompanying CO loss. The product can be further expanded to a decarhodium cage as shown in Eqn 2.24. The structure of $[Rh_{10}(CO)_{21}]^{2-}$ consists of a trigonal bipyramid and two face-sharing octahedra, and comparison of Figs. 2.30a and 2.32a shows that this Rh_{10}-cluster retains the Rh_9-motif of $[Rh_9(CO)_{19}]^{3-}$ as an integral part of the metal-framework. When treated with halide ion, $[Rh_{10}(CO)_{21}]^{2-}$ is degraded to $[Rh_9(CO)_{19}]^{3-}$.

$$[Rh_9(CO)_{19}]^{3-} \xrightarrow[\text{– 2MeCN}]{[Rh(CO)_2(NCMe)_2]^+} [Rh_{10}(CO)_{21}]^{2-} \qquad \textbf{Eqn 2.24}$$

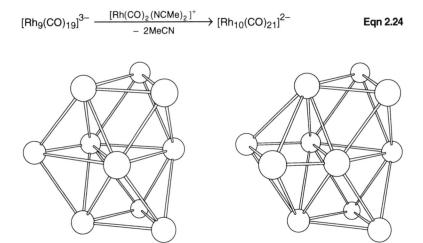

(a)   (b)

Fig. 2.32 The solid-state structures of the rhodium cores of (a) $[Rh_{10}(CO)_{21}]^{2-}$ and (b) $[Rh_{11}(CO)_{23}]^{3-}$.

Further expansion of the rhodium cage may be achieved by fragment condensation (Eqn 2.25) and the structural relationship between the precursor (Fig. 2.32a) and product (Fig. 2.32b) is clear. The insertion of the additional Rh atom into the metal cage results in a Rh_{11}-framework consisting of three face-sharing octahedra.

$$[Rh_{10}(CO)_{21}]^{2-} \xrightarrow[\text{– 2CO}]{[Rh(CO)_4]^-} [Rh_{11}(CO)_{23}]^{3-} \qquad \textbf{Eqn 2.25}$$

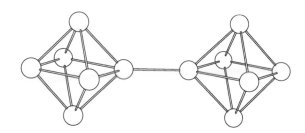

Fig. 2.33 The solid state structure of the metal core of $[Rh_{12}(CO)_{30}]^{2-}$.

A range of larger rhodium clusters is known. The sodium salt of $[Rh_{12}(CO)_{30}]^{2-}$ can be prepared from $Rh_4(CO)_{12}$ by reaction with sodium acetate in water/acetone, followed by treatment with NaCl. The structure of $[Rh_{12}(CO)_{30}]^{2-}$ consists of two octahedral sub-clusters, linked by a direct Rh–Rh bond (Fig. 2.33). In contrast, the iridium anions $[Ir_{12}(CO)_{26}]^{2-}$ and $[Ir_{12}(CO)_{24}]^{2-}$ (prepared according to Eqns 2.26 and 2.27) have quite different structures as shown in Figs. 2.34 and 2.35 respectively, although again, the structures feature octahedral building-blocks. In $[Ir_{12}(CO)_{26}]^{2-}$,

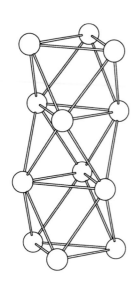

Fig. 2.34 The solid state structure of the metal core of $[Ir_{12}(CO)_{26}]^{2-}$.

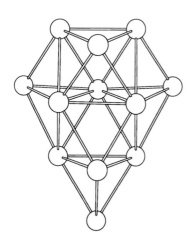

Fig. 2.35 The solid state structure of the metal core of $[Ir_{12}(CO)_{24}]^{2-}$.

there are three face-sharing octahedra, the origins of which can be viewed in terms of the condensation of the two Ir_6-octahedra of the precursor $[Ir_6(CO)_{15}]^{2-}$ — the fusion generates a third octahedral unit between the first two. The condensed polyhedral Ir_{12}-core in $[Ir_{12}(CO)_{24}]^{2-}$ consists of two octahedra, two trigonal bipyramids and one tetrahedron. Significantly, the same arrangement of atoms observed in $[Rh_{10}(CO)_{21}]^{2-}$ (Fig. 2.32a) forms part of the structure of $[Ir_{12}(CO)_{24}]^{2-}$.

$$2[Ir_6(CO)_{15}]^{2-} + 2[Cu(NCMe)_4]^+ \rightarrow [Ir_{12}(CO)_{26}]^{2-} + 2Cu + 4CO + 8MeCN$$

Eqn 2.26

$$[Ir_6(CO)_{15}]^{2-} \xrightarrow[\text{30 min, room temp.}]{[Cu(NCMe)_4]^+, \text{ THF}} \xrightarrow[]{\text{reflux, 12 hr.}} [Ir_{12}(CO)_{24}]^{2-}$$

Eqn 2.27

As the cluster nuclearity increases, it becomes increasingly clear that the metal atoms are arranged in such a way that the cluster core is beginning to resemble a fragment of a bulk metal lattice. This is even more apparent in clusters such as $[Rh_{14}(CO)_{25}]^{4-}$, $[Rh_{14}(CO_{26}]^{2-}$ and $[Rh_{15}(CO)_{30}]^{2-}$, the structures of which have been likened to body-centred cubic lattices, and in $[Rh_{15}(CO)_{27}]^{3-}$ and $[Rh_{17}(CO)_{30}]^{3-}$ which possess metal cores resembling fragments of hexagonal closed-packed lattices. The core of $[Rh_{22}(CO)_{37}]^{4-}$ consists of four close packed layers in a 3:6:7:6 stack. The selective preparation of these species is not easy. Equation 2.28 shows a route which can be pushed in favour of one particular product by changing the reaction time; for example, $\approx$ 30 min. leads to a good yield of $[Rh_{15}(CO)_{27}]^{3-}$ but after 3 hr, $[Rh_{14}(CO)_{25}]^{4-}$ is the main product.

$$Rh(CO)_2(acac) + Cs[PhCO_2] \xrightarrow[\text{glyme}]{CO/H_2, \text{ 15 atm, 410 K}} [Rh_{15}(CO)_{27}]^{3-} +$$

$$[Rh_{14}(CO)_{25}]^{4-} + [H_2Rh_{13}(CO)_{24}]^{3-}$$

Eqn 2.28

Several large osmium clusters also possess metal cores with structures that resemble pieces of close-packed metal lattices, and one of these is the dianion $[Os_{20}(CO)_{40}]^{2-}$, prepared as shown in Eqn 2.29. Figure 2.36 illustrates that the metal atoms in $[Os_{20}(CO)_{40}]^{2-}$ are in a cubic close-packed arrangement.

$$Os_3(CO)_{10}(NCMe)_2 \xrightarrow[\text{2. } [(Ph_3P)_2N]Cl \text{ in acetone / methanol at reflux}]{\text{1. vacuum pyrolysis, 570 K}} [Os_{17}(CO)_{36}]^{2-} +$$

$$[Os_{20}(CO)_{40}]^{2-} + [Os_{10}(CO)_{24}C]^{2-}$$

$$\text{as } [(Ph_3P)_2N]^+ \text{ salts} \qquad \textbf{Eqn 2.29}$$

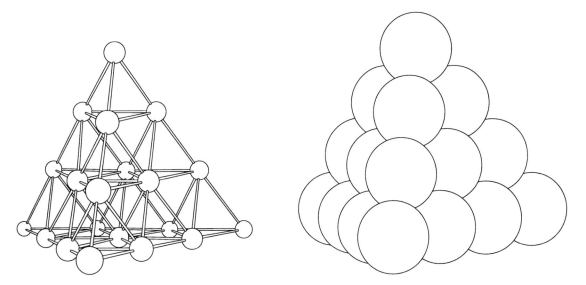

Fig. 2.36 The crystallographically confirmed Os_{20} core of $[Os_{20}(CO)_{40}]^{2-}$. The atoms are arranged in a ccp manner as the space-filling diagram on the right-hand side emphasizes.

Even larger are the platinum-containing cluster anions $[Pt_{24}(CO)_{30}]^{2-}$ $[Pt_{26}(CO)_{32}]^{2-}$, $[Pt_{38}(CO)_{44}]^{2-}$ and $[Ni_{38}Pt_6(CO)_{48}H_n]^{2-}$ (n = 0-3). The first two anions can be prepared from $[Pt_{15}(CO)_{30}]^{2-}$ (Fig. 2.31) in THF at reflux; $[Pt_{24}(CO)_{30}]^{2-}$ possesses a cubic close-packed array of platinum atoms, whilst in $[Pt_{26}(CO)_{32}]^{2-}$ the metal atoms are hexagonal close-packed. The stacked cluster anion $[Pt_9(CO)_{18}]^{2-}$ is the precursor to $[Pt_{38}(CO)_{44}]^{2-}$ (Eqn 2.30) in which the platinum atoms adopt a cubic close-packed arrangement; the metal core is large enough that six Pt atoms are interstitial — a situation that gives these atoms a truly bulk metal 'feel'.

$$[AsPh_4]_2[Pt_9(CO)_{18}] \xrightarrow[\text{2. } CF_3CO_2H]{\text{1. MeCN, reflux}} [AsPh_4]_2[Pt_{38}(CO)_{44}] \qquad \textbf{Eqn 2.30}$$

3 Bonding schemes for transition metal clusters

We have already seen that in dimetallic systems and in *small* clusters, each transition metal centre may be expected to obey the 18-electron rule (with some exceptions, e.g. for group 9 and 10 metals which may possess 16-electron configurations). However, as the nuclearity of the cluster increases, this type of approach to the bonding becomes inconvenient, and in this chapter, we consider bonding schemes which treat the polyhedral skeleton as a single unit or as an assembly of condensed units requiring a defined number of cluster bonding electrons.

3.1 Metal carbonyl clusters as analogues of boranes

3.1.1 Wade's rules: an overview

A *deltahedron* is a polyhedron with triangular faces.
e.g. An octahedron is a deltahedron, but a trigonal prism is not.

In the accompanying text in this series, *Cluster Molecules of the p-Block Elements*, we introduced the Polyhedral Skeletal Electron Pair Theory (*PSEPT*) (or Wade's rules) and used it to rationalize the bonding in boranes and related molecules including metallaboranes. Within this method, clusters are described as being *closo* if the core-geometry consists of a closed deltahedron, and these *closo* cages are the 'parents' from which other geometries are derived. Each derivative cage is related to the parent by the removal of one, two or three (and so on) vertices as shown in Fig. 3.1.

Full details of this method of electron counting will not be discussed here, and the reader is guided to the companion Primer *Cluster Molecules of the p-Block Elements*, C.E. Housecroft, 1994.

The hydroborate dianion $[B_6H_6]^{2-}$ is a *closo*-cluster, and B_5H_9 and B_4H_{10} are *nido*- and *arachno*-species which possess the same number of cluster bonding electrons as $[B_6H_6]^{2-}$. Each BH unit provides two electrons and each additional H atom provides one electron, making each of $[B_6H_6]^{2-}$, B_5H_9 and B_4H_{10} a 14-electron (i.e. 7 electron pair) cluster.

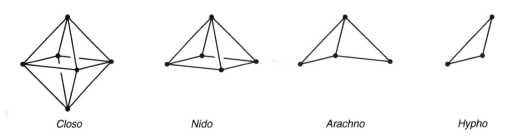

| Closo | Nido | Arachno | Hypho |

Fig. 3.1 The relationship between *closo-*, *nido-*, *arachno-* and *hypho*-clusters, exemplified by the octahedral cage.

3.1.2 Use of the isolobal principle and cluster-fragment electron counts

By using the isolobal principle, analogies can be drawn between some transition metal carbonyl and borane clusters. For example, BH and C_{3v} $Ru(CO)_3$ fragments are isolobal (Fig. 3.2), and the anion $[Ru_6(CO)_{18}]^{2-}$ is an analogue of $[B_6H_6]^{2-}$. The respective electron counts are as follows:

$[B_6H_6]^{2-}$ $[Ru_6(CO)_{18}]^{2-}$

6 BH = 6 × 2 electrons   6 $Ru(CO)_3$ = 6 × 2 electrons
2– charge = 2 electrons   2– charge = 2 electrons
Total count = 14 electrons   Total count = 14 electrons
= 7 pairs   = 7 pairs

Each dianion therefore has seven pairs of electrons with which to bond six skeletal atoms. This is consistent with the adoption of a *closo*-structure, and both $[B_6H_6]^{2-}$ and $[Ru_6(CO)_{18}]^{2-}$ possess octahedral cages. It must be stressed that the application of *PSEPT* to metal carbonyl clusters provides a rationalization of the geometry of the *cluster core* but it does *not* provide information about the disposition of the CO ligands.

For the Wade approach, the number of electrons that a metal-based fragment provides for cluster bonding can be determined using Eqn 3.1.

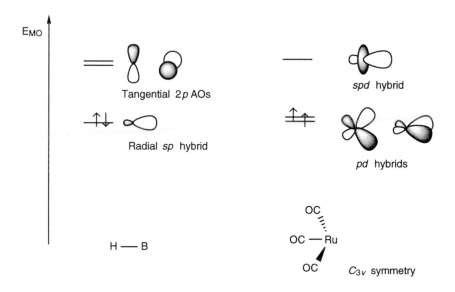

Fig. 3.2 The frontier orbitals of a BH unit consist of a radial and two tangential orbitals. (The BH σ and σ* MOs are not shown in the diagram). A C_{3v} $Ru(CO)_3$ unit is isolobal with a BH unit — there are the same number of frontier orbitals with the same symmetries and approximate energies. The *order* of the MOs does not influence the bonding capabilities of the fragments. Each fragment has two electrons (those in the frontier MOs) which are available for cluster bonding.

> If x = number of valence electrons provided by a transition metal fragment for cluster bonding, then
>
> $$x = v + n - 12 \qquad\qquad \textbf{Eqn 3.1}$$
>
> where v = number of valence electrons from the metal atom and n = number of electrons donated by the ligands.

Thus, for a $Ru(CO)_3$ unit, $v = 8$, $n = (3 \times 2) = 6$, and $x = 2$. Similarly, $x = 2$ for $Fe(CO)_3$ and $Os(CO)_3$ units. Table 3.1 emphasizes the relationship between the number of electrons available and the position of the metal in the d-block. This method of electron counting is not restricted to carbonyl ligands; for example, a phosphine ligand is a 2-electron donor, an η^5-C_5H_5 ligand is a 5-electron donor, and an η^6-C_6H_6 ligand is a 6-electron donor. Interstitial atoms provide all their valence electrons to cluster bonding; for example, B provides three electrons, C provides four electrons, and N gives five. Examples of these and other systems are given in later chapters.

With regard to Table 3.1, the reader should consider the consequences on the bonding in a cluster of the involvement of fragments which have *negative* values of x.

Table 3.1 The number of electrons, x, available from metal carbonyl units for cluster bonding within the Wade approach. Values of x are determined using Eqn 3.1

	group 6 Cr, Mo, W	group 7 Mn, Tc, Re	group 8 Fe, Ru, Os	group 9 Co, Rh, Ir	group 10 Ni, Pd, Pt
$M(CO)_2$	−2	−1	0	1	2
$M(CO)_3$	0	1	2	3	4
$M(CO)_4$	2	3	4	5	6

3.1.3 Structural predictions and rationalizations

The general strategy for rationalizing a geometry or predicting a metal-core structure is set out below as a series of guidelines. This is followed by some examples of the application of Wade's rules to transition metal carbonyl clusters.

A set of guidelines may be followed in order to rationalize the structure of the core of a metal carbonyl cluster

1. How many $M(CO)_4$, $M(CO)_3$ or $M(CO)_2$ units are there?
2. How many additional CO ligands are there?
3. How many valence electrons do these units provide for cluster bonding?
4. Is there an overall charge? If so, how many electrons does this provide to (if an anion) or remove from (if a cation) the total electron count?
5. What is the parent deltahedron? — n vertices requires $(n+1)$ electron pairs.
6. After each metal centre has been accommodated at a skeletal vertex, are there any vertices vacant in the parent deltahedron? What class is the cluster — *closo, nido, arachno* or *hypho* ?
7. It is *not* possible to state with any certainty where the carbonyl ligands will reside.

These guidelines can be extended to include other ligands as illustrated in later chapters.

Example 1: $Os_5(CO)_{16}$

Rationalize why the metal core-structure of $Os_5(CO)_{16}$ is a closed trigonal bipyramid.

The formula $Os_5(CO)_{16}$ may be broken down as follows:

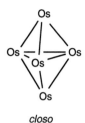

closo

5 $Os(CO)_3$ units	= 5 × 2 electrons
1 additional CO	= 2 electrons
Total electron count	= 12 electrons = 6 electron pairs

Six electron pairs are consistent with a parent polyhedron with 5 vertices, i.e. a trigonal bipyramid.

NB: The partitioning of the formula into the $Os(CO)_3$ units and one CO is a formalism — it does not necessarily mean that the ligands are partitioned in this manner in the molecule. The reader should try several other ways of breaking down the formula, and show that it makes no difference to the final result.

Example 2: $Ir_4(CO)_{12}$

Predict the geometry of the metal core of $Ir_4(CO)_{12}$.

The formula $Ir_4(CO)_{12}$ may be partitioned as follows:

4 $Ir(CO)_3$ units	= 4 × 3 electrons
Total electron count	= 12 electrons = 6 electron pairs

Six electron pairs are consistent with a parent polyhedron with 5 vertices, i.e. a trigonal bipyramid. There are only 4 Ir atoms to be accommodated and so the cluster has a *nido* structure — a tetrahedron.

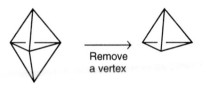

Remove a vertex

nido

Example 3: $[Co_6(CO)_{15}]^{2-}$

Predict the geometry of the metal core of $[Co_6(CO)_{15}]^{2-}$.

The formula $[Co_6(CO)_{15}]^{2-}$ may be partitioned as follows:

6 $Co(CO)_2$ units	= 6 × 1 electrons
3 additional CO ligands	= 3 × 2 electrons
2– charge	= 2 electrons
Total electron count	= 14 electrons = 7 electron pairs

Seven electron pairs are consistent with a parent polyhedron with 6 vertices, i.e. an octahedron and so $[Co_6(CO)_{15}]^{2-}$ is predicted to have an octahedral core.

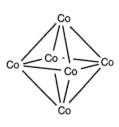

closo

The approach outlined above can be applied successfully to a range of relatively small metal clusters but there are exceptions. For example, the method correctly predicts that $[Ni_6(CO)_{12}]^{2-}$ possesses an octahedral structure, but fails to show that $[Pt_6(CO)_{12}]^{2-}$ has a trigonal prismatic geometry.

3.1.4   The capping principle

The justification for the capping principle lies in the fact that the frontier orbitals of the capping unit can interact with orbitals that point out from the M_3-face to be capped, and these interactions do not disrupt the bonding in the cluster-core.

Earlier in this chapter, we saw that condensed polyhedral structures are prevalent amongst transition metal carbonyl clusters. For example, $Os_7(CO)_{21}$ adopts a monocapped octahedral structure (Fig. 2.27). This is classified as a *capped-closo* geometry and the electron count is as follows:

$$7 \; Os(CO)_3 \text{ units} \quad = 7 \times 2 \text{ electrons}$$
$$\text{Total electron count} \quad = 14 \text{ electrons} = 7 \text{ electron pairs}$$

There are 7 electron pairs with which to bond 7 metal centres. If there were only 6 metal centres, the structure would be a *closo* cage; by Wade's rules, *the addition of a capping unit requires no further electrons*. Similarly, $[Os_8(CO)_{22}]^{2-}$ possesses 7 electron pairs for cluster bonding and adopts a bicapped octahedral structure (Fig. 3.3).

Fig. 3.3   *Closo* and *capped closo* structures require the same number of electrons within Wade's rules. In this example, the 6-vertex octahedron, the 7-vertex monocapped octahedron, and the 8-vertex bicapped octahedron all require 7 electron pairs.

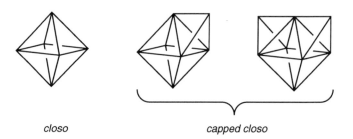

closo   *capped closo*

Further application of the capping principle shows that n atoms with $(n+1)$ pairs of cluster bonding electrons do not necessarily have to adopt a *closo*-structure but could instead form a *capped-nido* geometry. Thus, the *closo*-octahedron and the *capped-nido* square-based pyramid both require 7 pairs of cluster bonding electrons (Fig. 3.4). In practice, *closo*-structures are more common, but an example of a *capped-nido* geometry is $H_2Os_6(CO)_{18}$.

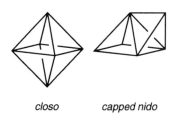

closo   *capped nido*

Fig. 3.4   A cluster with six metal atoms and seven pairs of bonding electrons could, by Wade's rules, adopt either a *closo*-octahedral or a *capped-nido* square-based pyramidal structure. In practice, *closo*-structures are more common.

3.1.5   Summary

Wade's approach can be used to rationalize the structures of transition metal carbonyl (and related) clusters that are analogues of boranes and derived anions. The method can be extended to capped structures, but begins to show limitations as the nuclearity of the metal core increases. An important difference between boranes and transition metal clusters is the propensity for the former to adopt large cages with a 'hole' inside, such as those observed in $B_{10}H_{14}$ and $[B_{12}H_{12}]^{2-}$, whereas high nuclearity metal cages tend to be constructed from smaller, condensed units such as tetrahedra and octahedra.

3.2 Total valence electron counts for simple structures

An alternative method of rationalizing the geometries of transition clusters, which is still within the remit of *PSEPT*, is to consider the total number of valence electrons that are available for bonding. We begin by considering some common and relatively simple metal cage structures, and in the next section discuss more complex geometries.

Each particular transition metal cluster geometry has associated with it a characteristic number of valence electrons. Some of these electron counts are summarized in Table 3.2, and their application is illustrated in the worked examples that follow.

It is beyond the scope of this text to consider the origins of the numbers of electrons that are characteristic of a particular geometry. Suggestions for further reading are given on p. 4.

Table 3.2 Characteristic total valence electron counts for selected transition metal clusters (see Fig. 3.5).

Cluster cage geometry	Valence electron count
Triangle	48
Tetrahedron	60
Butterfly or a planar, 4-atom raft	62
Square	64
Trigonal bipyramid	72
Square-based pyramid	74
Octahedron	86
Trigonal prism	90

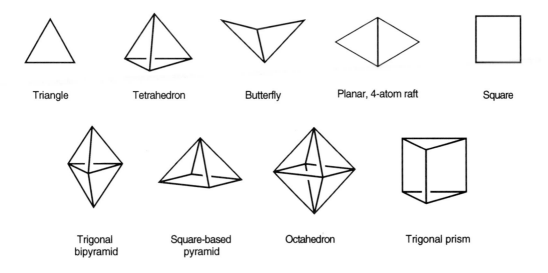

Fig. 3.5 Some cage geometries found for transition metal carbonyl clusters with six or less atoms.

Example 1: [Co₃(CO)₁₀]⁻

Is the triangular core of [Co₃(CO)₁₀]⁻ consistent with the electron count?

The total number of valence electrons available is:

$$3 \; Co^0 \; [Ar]4s^23d^7 \quad = 3 \times 9 = 27 \text{ electrons}$$
$$10 \; CO \qquad\qquad = 20 \text{ electrons}$$
$$1- \text{charge} \qquad = 1 \text{ electron}$$
$$\text{Total valence electron count} \quad = 48 \text{ electrons}$$

A count of 48 electrons is consistent with a triangular structure.

This method of counting electrons is more convenient than applying the 18-electron rule to each metal centre as shown in Section 2.3.2. As with Wade's rules, the result only rationalizes the geometry of the metal core, and does *not* provide information about the disposition of the ligands.

Example 2: Os₅(CO)₁₆

Rationalize why the core-structure of Os₅(CO)₁₆ is a trigonal bipyramid.
The total number of valence electrons available is:

$$5 \; Os^0 \; [Xe]6s^25d^6 \quad = 5 \times 8 = 40 \text{ electrons}$$
$$16 \; CO \qquad\qquad = 32 \text{ electrons}$$
$$\text{Total valence electron count} \quad = 72 \text{ electrons}$$

72 electrons is consistent with a trigonal bipyramidal structure.

Example 3: Ir₆(CO)₁₆

Predict the structure of the metal core in Ir₆(CO)₁₆.
The total number of valence electrons available is:

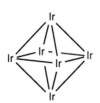

$$6 \; Ir^0 \; [Xe]6s^25d^7 \quad = 6 \times 9 = 54 \text{ electrons}$$
$$16 \; CO \qquad\qquad = 32 \text{ electrons}$$
$$\text{Total valence electron count} \quad = 86 \text{ electrons}$$

A count of 86 electrons is consistent with an octahedral structure.

The structures of Os₄(CO)₁₄, Os₄(CO)₁₅ and Os₄(CO)₁₆ were described in Section 2.4.

Note that in Table 3.2 there are several sets of structures, the electron counts for which differ by 2 or 4 electrons. Figure 3.6 illustrates one group of structures, and shows that the *addition of a pair of electrons opens up one edge of the metal skeleton.* Series of related clusters are known, members of which differ by one 2-electron ligand, e.g. Os₄(CO)₁₄, Os₄(CO)₁₅ and Os₄(CO)₁₆, but in practice, the interconversion of such species may not always be achieved simply. Nonetheless, total valence electron counts can provide a means of predicting the outcome of a 2-electron reduction or oxidation, whether this is carried out electrochemically or by the addition or elimination of a 2-electron donor ligand, or by some other means.

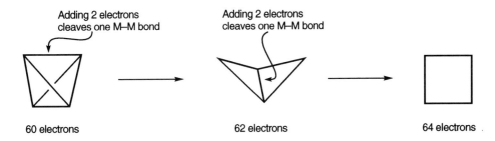

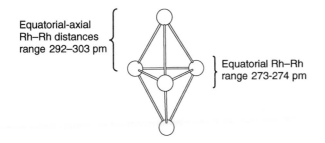

Fig. 3.6 The addition of a pair of electrons opens up one edge of the metal cage. The step from the 62-electron butterfly to the 64-electron square is one of two possibilities. What other structure would be consistent with 64 electrons?

The electron counts given in Table 3.2 are not without exceptions. For example, $[Rh_5(CO)_{15}]^-$ has a 76 valence electron count, and yet adopts a trigonal bipyramidal structure, the expected count for which is 72. In fact, crystallographic data for $[Rh_5(CO)_{15}]^-$ reveal that the Rh–Rh distances between the equatorial and axial atoms are greater than between equatorial metal centres (Fig. 3.7). This is in keeping with the notion that the addition of a pair of electrons cleaves an M–M edge — however, in this case, and in a range of other clusters, additional electrons cause *bond lengthening* rather than bond cleavage.

Fig. 3.7 The variation in Rh–Rh distances in $[Rh_5(CO)_{15}]^-$, a 76-electron cluster.

3.3 Total valence electron counts for condensed clusters

Clusters with condensed polyhedral structures can be considered in terms of sub-cluster units which share one or more:

> metal atoms (e.g. $Os_5(CO)_{19}$, Fig. 2.23)
> M–M edges (e.g. $Os_5(CO)_{18}$, Fig. 2.22)
> M_3-triangular faces (e.g. $Os_7(CO)_{21}$, Fig. 2.27).

The number of valence electrons required by a particular condensed polyhedral structure is equal to the number of electrons required by each sub-cluster, minus the electron count for the shared unit.

Subtract *18 electrons for a shared metal atom,*
34 electrons for a shared edge, or
48 electrons for a shared triangular face.

Example 1: Os$_5$(CO)$_{19}$

Os$_5$(CO)$_{19}$ has a 'bow-tie' geometry. Is this consistent with the number of valence electrons available?

The problem needs to be approached in two parts: counting the number of available electrons, and determining the electrons required by the structure.

The total number of valence electrons available in Os$_5$(CO)$_{19}$ is:

$$5 \text{ Os}^0 \ [Xe]6s^25d^6 \quad = 5 \times 8 = 40 \text{ electrons}$$
$$19 \text{ CO} \quad = 38 \text{ electrons}$$
$$\text{Total valence electron count} \quad = 78 \text{ electrons}$$

Now consider how the cage can be partitioned into sub-units:

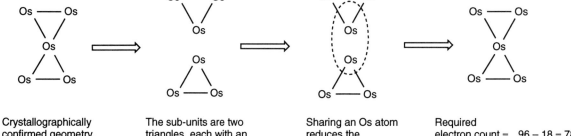

Crystallographically confirmed geometry.

The sub-units are two triangles, each with an electron count of 48. The total electron count for two triangles is 96.

Sharing an Os atom reduces the total electron count by 18.

Required electron count = 96 – 18 = 78

The observed structure for Os$_5$(CO)$_{19}$ is therefore consistent with the total number of valence electrons available.

Example 2: Os$_7$(CO)$_{21}$

Confirm that the monocapped octahedral structure of Os$_7$(CO)$_{21}$ is consistent with the number of valence electrons available.

The cage can be partitioned into two skeletal units, which share a common Os$_3$-triangle:

Crystallographically confirmed geometry.

Octahedron Electron count = 86 + Tetrahedron Electron count = 60

Total electron count for the monocapped octahedron = 86 + 60 – 48 = 98

The total number of valence electrons available is:

$$7 \text{ Os}^0 \ [Xe]6s^25d^6 \quad = 7 \times 8 = 56 \text{ electrons}$$
$$21 \text{ CO} \quad = 42 \text{ electrons}$$
$$\text{Total valence electron count} \quad = 98 \text{ electrons}$$

The observed structure for Os$_7$(CO)$_{21}$ is consistent with the total number of valence electrons available.

Example 3: $Os_5(CO)_{18}$

$Os_5(CO)_{18}$ is a raft-cluster. Is this consistent with the number of valence electrons available?

The total number of valence electrons available is:

$$5\ Os^0\ [Xe]6s^2 5d^6\ = 5 \times 8 = 40\ \text{electrons}$$
$$18\ CO\ = 36\ \text{electrons}$$
$$\text{Total valence electron count}\ = 76\ \text{electrons}$$

The structure of $Os_5(CO)_{18}$ (Fig. 2.22) is constructed from three triangles, which share two common edges:

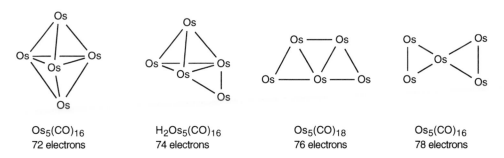

Crystallographically confirmed geometry.

Electron count for three triangles= 3 x 48 = 144

Total electron count for the raft = 144 – (2 x 34) = 76

The relationship between the structures of $Os_5(CO)_{18}$ and $Os_5(CO)_{19}$ reflects a difference in total electron counts of two — two electrons that are provided by a CO ligand. The series can be further extended as shown in Fig. 3.7; each cluster bound H atom in $H_2Os_5(CO)_{16}$ contributes one electron to cluster bonding.

$Os_5(CO)_{16}$	$H_2Os_5(CO)_{16}$	$Os_5(CO)_{18}$	$Os_5(CO)_{16}$
72 electrons	74 electrons	76 electrons	78 electrons

Fig. 3.7 Structural relationships and total valence electron counts for a series of pentaosmium clusters. Which Os–Os bonds have been broken in going from one Os_5-core to the next? Can any other structural isomers be predicted which retain the same electron counts?

For some high nuclearity clusters such as the large rhodium species described earlier in this chapter, the number of available electrons may differ from the number calculated by the method described here. For example, the $[Rh_9(CO)_{19}]^{3-}$ anion has 122 electrons available, but adopts the bi-octahedral structure shown in Fig. 2.30a, the electron count for which is 124. For detailed discussions on this matter, the reader is guided to the works of Mingos (see *Further Reading*, p. 4).

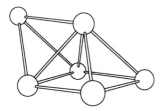

Fig. 3.8 The Os_6-core structure of $H_2Os_6(CO)_{18}$.

Clusters with interstitial atoms are described in Chapter 7.

^{11}B and ^{103}Rh are spin active nuclei:

^{11}B $(I = ^3/_2; 80\%)$

^{103}Rh $(I = ^1/_2; 100\%)$

3.4 Electron counting schemes in the context of experimental results

In the preceding discussion, we have illustrated how metal-core geometries are dependent upon the number of valence electrons required. The examples given have generally asked for a rationalization of a crystallographically determined structure. Working from crystallographic data means that we have a clear picture of the cluster molecule or ion in the *solid state*.

In practice, new compounds do not always form X-ray quality crystals, and the experimentalist may have formulated a compound based on elemental analysis, mass spectrometric, and IR and NMR spectroscopic data. Knowing a formula allows the number of valence electrons available to be determined, and with a knowledge of the number of metal atoms and of related compounds, predictions can often be made concerning structure. An 86-electron cluster is very likely to have an octahedral metal core, and a 60-electron cluster is usually found to be tetrahedral. But this is not always the case. The compound $H_2Os_6(CO)_{18}$ has an 86-electron count — an octahedron might be predicted, but crystallographic results reveal it to be a capped square-based pyramid (Fig. 3.8), a geometry that also requires 86 electrons.

As the nuclearity of the cluster increases, it becomes more difficult to predict the core-structure of a new type of species with any certainty and single crystal X-ray crystallographic data are essential. However, in solution, skeletal isomerism is possible and so it is important not to assume that the crystallographic data tell the whole structural story. For example, the anion $[Rh_2Ru_4(CO)_{16}B]^-$ possesses an octahedral metal cage enclosing an interstitial boron atom. The results of an X-ray diffraction study of $[(Ph_3P)_2N][Rh_2Ru_4(CO)_{16}B]$ confirm that the rhodium atoms are mutually *trans*. However, the solution ^{11}B NMR spectrum of the same salt exhibits two signals (δ +197 and +194), both triplets (J_{Rh-B} 26 Hz) and these are assigned to the *trans* and *cis* isomers of the cluster (Fig. 3.9). Both isomers meet the requirements of an 86-electron cluster.

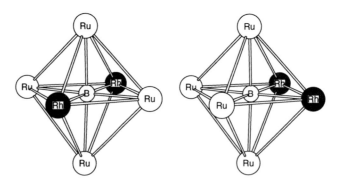

Fig. 3.9 The boron-containing metal cores of *trans*-$[Rh_2Ru_4(CO)_{16}B]^-$ and *cis*-$[Rh_2Ru_4(CO)_{16}B]^-$.

4 Hydride ligands

The aims of this and the following chapters are to introduce aspects of the chemistry of transition metal carbonyl dimers and clusters, and, at the same time, illustrate classes of compound such as those with hydride ligands, phosphine and phosphite ligands, organic ligands, interstitial atoms, and other groups involving elements from the p-block. We shall also consider selected methods to prepare heterometallic clusters.

In the space available, it is only possible to provide the reader with an introduction to the area; several books and reviews are recommended for more detailed accounts (*Further reading*, p. 4).

4.1 Bonding and spectroscopy

Many dimers and clusters contain hydrogen atoms attached directly to one or more metal centres, and such ligands are generally termed *hydrides*. This implies a δ^- charge, (the metal being more electropositive than the hydrogen atom), but in many systems, the electron withdrawing properties of the other ligands in the molecule mean that the metal-bound hydrogen atom can be removed by treatment with base, or introduced by reaction of an anionic dimer or cluster with acid (Eqn 4.1). Examples of the introduction (or removal) of hydrogen ligands both in the form of H^+ and H^- are given in this chapter.

$$[Fe_2(CO)_8]^{2-} + H^+ \rightarrow [HFe_2(CO)_8]^- \qquad \textbf{Eqn 4.1}$$

The four general modes of attachment of a hydride ligand to a metal dimer or cluster are illustrated in Fig. 4.1. When terminally attached, the M–H bond can be considered to be a localized 2-centre 2-electron interaction, but delocalized descriptions may be invoked for the other modes. The M-H-M bridge may be considered in terms of a 3-centre 2-electron interaction, with the $1s$ atomic orbital of the hydrogen atom overlapping with suitable atomic or hybrid orbitals on the metal centres (Fig. 4.2). Similarly, for the face-capping and interstitial hydrides, multi-centred bonding descriptions are required. The 3-centre 2-electron bonding description for the M-H-M interaction is in keeping with the experimental observation that, in general, a hydride-bridged metal–metal bond tends to be longer than a corresponding unbridged bond.

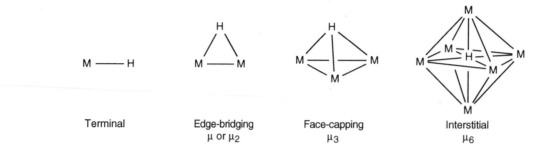

Terminal   Edge-bridging   Face-capping   Interstitial
            μ or μ₂   μ₃   μ₆

Fig. 4.1   The principal modes of attachment of hydride ligands to metal centres. The interstitial mode is shown specifically for an octahedral cluster.

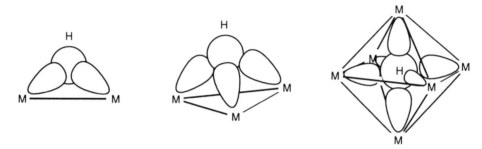

Fig. 4.2   Overlap of the H 1s AO with suitable atomic or hybrid orbitals on the metal centres in bridging, face-capping and interstitial modes of attachment. The metal-hydrogen bonding in each case is delocalized.

The presence of a metal hydride ligand can be detected by using IR or NMR spectroscopy, the latter being more useful as a diagnostic tool in the laboratory. Infrared absorptions due to the ν(M-H) mode are often weak. In the ^{1}H NMR spectrum, terminal, μ- and μ₃-hydrides are often characterized by highfield (low frequency) signals due to the fact that the hydrides are significantly shielded; the *approximate* range is δ −8 to −30. The chemical shifts of *interstitial* hydrides vary a great deal and may occur at highfield or at unusually high frequency (lowfield). In the octahedral clusters [(μ₆-H)Co₆(CO)₁₅]⁻ and [(μ₆-H)Ru₆(CO)₁₈]⁻, signals due to the hydrides are observed at δ +23.2 and +16.4 respectively.

Hydride resonances are usually sharp, *but* a word of caution concerning their intensities is needed — the relative intensity of a hydride signal with respect to, say, organic protons is generally very low. For example, the relative integrals of the phenyl : hydride signals in the ^{1}H NMR spectrum of the salt [(Ph₃P)₂N][HFe₃(CO)₁₁] is theoretically   30 : 1. In practice, integrals close to 350 : 1 may be observed due to the differing relaxation times of the organic and metal-bound protons.

Spin-spin coupling between ^{1}H nuclei in different environments, and ^{1}H and heteronuclei such as ^{31}P (100% $I = {}^1/_2$) and ^{103}Rh (100% $I = {}^1/_2$), can provide valuable structural information about the position of a hydride ligand in a molecule. Consider the cation shown in Fig. 4.3. The hydride ligand symmetrically bridges the Ru–Ru bond and couples to four equivalent ^{31}P nuclei; a binomial (1:4:6:4:1) quintet is observed, centred at δ –10.2. The coupling constant of J_{P-H} 10.8 Hz is typical of a *cis*-relationship between ^{31}P and ^{1}H nuclei.

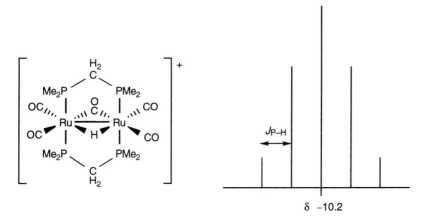

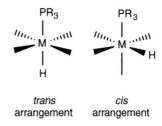

Fig. 4.3 The structure of $[Ru_2(CO)_4(\mu\text{-}CO)(\mu\text{-}Me_2PCH_2PMe_2)_2(\mu\text{-}H)]^+$ and the ^{1}H NMR spectroscopic signal for the hydride ligand.
The value of J_{P-H} can be measured between any pair of adjacent lines in the signal.

In general, values of $J_{P-H}(trans)$ are greater than $J_{P-H}(cis)$, typically 30 Hz versus 10-15 Hz (Fig. 4.4). Additionally, smaller long range [31]P-^{1}H spin-spin couplings may be observed.

There are many examples of cluster hydrides undergoing fluxional processes and low temperature ^{1}H NMR spectra may be required to obtain information about the static structure. Consider the anion $[H_3Ru_4(CO)_{12}]^-$ in which the ruthenium atoms form a tetrahedron. At 178 K, the ^{1}H NMR spectrum is consistent with the presence of two isomers labelled **A** and **B** in Fig. 4.5 — there are three signals, a singlet (δ –17.4, relative integral 3.9), a doublet (δ –15.9 J_{HH} 2.5 Hz, relative integral 2) and a triplet (δ –19.1 J_{HH} 2.5 Hz, relative integral 1). Above room temperature, the ^{1}H NMR spectrum exhibits a singlet at δ –16.9 which indicates that there is exchange of the cluster H atoms giving rise to one, average proton environment — this results in an interchange of isomers **A** and **B**. In some molecules, the activation barrier to fluxionality in solution is so low that the static structure is not frozen out at low temperature. Accessible experimental temperature ranges are affected by compound solubility and the temperature at which the solvent freezes.

Fig. 4.4 *Trans* and *cis* P-M-H arrangements in an octahedral geometry. Generally:

$$J_{PH(trans)} > J_{PH(cis)}$$

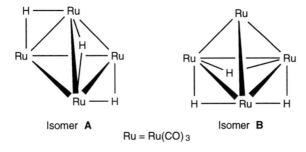

Isomer **A**   Isomer **B**

Ru = Ru(CO)$_3$

Fig. 4.5 Structural isomers of $[H_3Ru_4(CO)_{12}]^-$. In **A**, there are two hydride environments in a 1:2 ratio, and in **B** all three hydrides are equivalent.

40 *Hydride ligands*

4.2 Syntheses and structures of hydride-containing species

Anionic metal–metal bonded dimers may be protonated to give dimetallic products containing a hydride bridge. For example, $[Cr_2(CO)_{10}]^{2-}$ is a strong base and reacts with water according to Eqn 4.2. An alternative route to $[(\mu\text{-}H)Cr_2(CO)_{10}]^-$ is given in Eqn 4.3; the tetrahydroborate(1–) anion (often called borohydride) is a common source of hydride.

The tetrahydroborate(1–) anion $[BH_4]^-$ is commonly used as a source of hydride H^-.

$$[Cr_2(CO)_{10}]^{2-} + H_2O \rightarrow [(\mu\text{-}H)Cr_2(CO)_{10}]^- + [OH]^- \qquad \textbf{Eqn 4.2}$$

$$2Cr(CO)_6 \xrightarrow{Na[BH_4]} [(\mu\text{-}H)Cr_2(CO)_{10}]^- \qquad \textbf{Eqn 4.3}$$

The structure of the anion $[(\mu\text{-}H)Mo_2(CO)_{10}]^-$ is shown in Fig. 4.6; each metal centre is in an approximately octahedral environment, and the CO ligands adopt a staggered conformation. The Mo-H-Mo bridge is bent and the Mo–Mo distance (342 pm) appears to be too long for there to be a significant degree of direct metal–metal bonding; in $[Mo_2(CO)_{10}]^{2-}$, which has a metal–metal bond, the Mo–Mo distance is 312 pm. Thus, protonation has resulted in metal–metal bond cleavage, but the hydride bridge keeps the two metal fragments in close proximity.

Interestingly, the reported structures of several salts of $[(\mu\text{-}H)Cr_2(CO)_{10}]^-$ indicate that the geometry of this anion is cation-dependent. In $K[(\mu\text{-}H)Cr_2(CO)_{10}]$, the Cr---Cr separation in the anion is 336 pm (compared to the metal–metal bonded distance of 297 pm in $[Cr_2(CO)_{10}]^{2-}$) and the equatorial CO ligands are in an eclipsed conformation (Fig. 4.6). The hydride ligand lies midway along the Cr–Cr vector. In contrast, in the salt $[K(2,2,2\text{-crypt})][(\mu\text{-}H)Cr_2(CO)_{10}]$, the anion adopts a structure which is similar to that of the $[(\mu\text{-}H)Mo_2(CO)_{10}]^-$.

2,2,2-crypt is defined in Box 4.1.

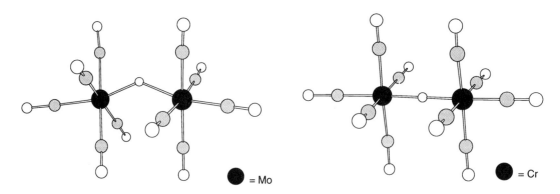

Fig. 4.6 The solid state structures of $[(\mu\text{-}H)Mo_2(CO)_{10}]^-$ in the salt $[(Ph_3P)_2N][(\mu\text{-}H)Mo_2(CO)_{10}]$, and of $[(\mu\text{-}H)Cr_2(CO)_{10}]^-$ in the salt $K[(\mu\text{-}H)Cr_2(CO)_{10}]$. The structure of $[(\mu\text{-}H)Cr_2(CO)_{10}]^-$ is cation dependent and in $[K(2,2,2\text{-crypt})][(\mu\text{-}H)Cr_2(CO)_{10}]$ the anion has a structure similar to that shown for $[(\mu\text{-}H)Mo_2(CO)_{10}]^-$.

Box 4.1 Counter-cations for large anions

For dimetallic and cluster anions, it is usually necessary to use a large counter-cation to stabilize and crystallize the product. Some common choices are the tetraphenylphosphonium ion $[PPh_4]^+$, the tetraphenylarsonium ion $[AsPh_4]^+$, and the bis(diphenylphosphine)nitrogen(1+) ion $[(Ph_3P)_2N]^+$ or $[PPN]^+$ (shown on the left below). Such large cations pack well with large anions in a crystal lattice as is exemplified in the diagram on the right-hand side below, which shows part of the packing diagram for $[(Ph_3P)_2N][(\mu\text{-}H)Fe_3(CO)_{11}]$ — the Fe_3-units are highlighted as triangles of black atoms.

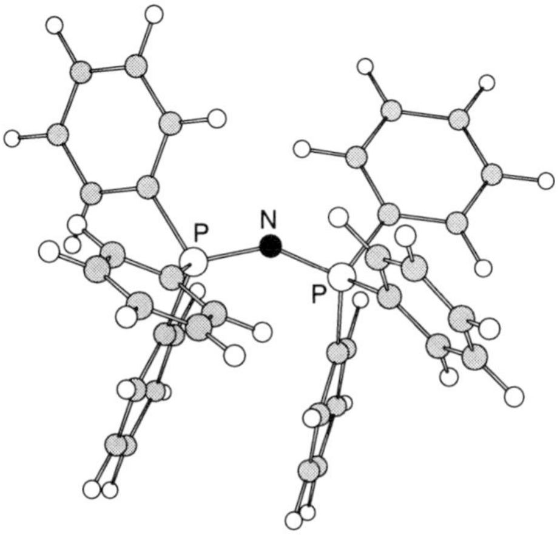

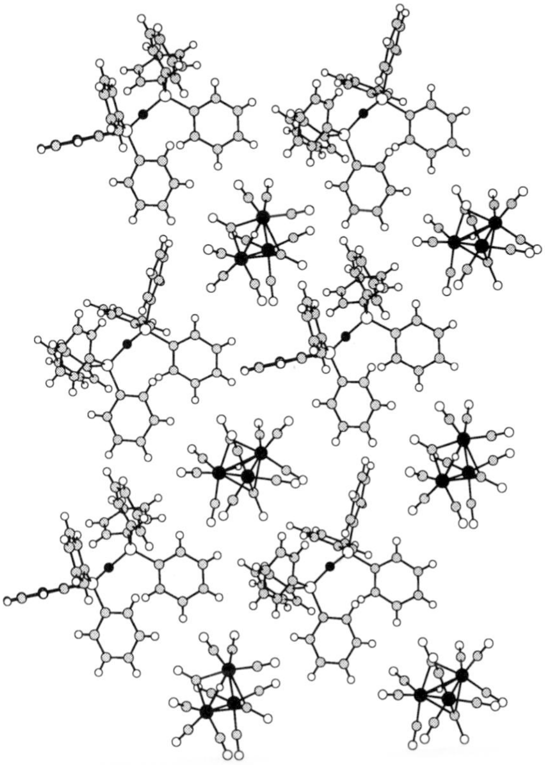

For alkali metal salts of dimetallic and cluster anions, it is common to use a macrocyclic ligand to complex with the group 1 metal ion; this generates a large complex-cation. In the example shown below, the ligand 2,2,2-cryptand (abbreviated to 2,2,2-crypt) coordinates to a K^+ ion and encapsulates the metal centre as the space-filling diagram illustrates.

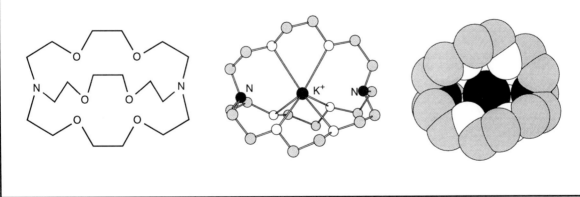

The protonation of $[Fe_2(CO)_8]^{2-}$ leads to the formation of $[HFe_2(CO)_8]^-$ and similarly, when $[Fe_3(CO)_{11}]^{2-}$ is treated with acetic acid, the anion $[HFe_3(CO)_{11}]^-$ is produced (Fig. 4.7). In both, the hydride ligand bridges an Fe–Fe edge, and a direct metal–metal bonding interaction is needed if each Fe centre is to obey the 18-electron rule. The Fe–Fe bond length is *shorter* in $[HFe_2(CO)_8]^-$ (252 pm) than in $[Fe_2(CO)_8]^{2-}$ (279 pm). The shortening of metal–metal edge when it is bridged by a hydride ligand is also observed in $[HFe_3(CO)_{11}]^-$; within the Fe$_3$-triangle, the H-bridged edge is 257 pm and the average length of the other two Fe–Fe bonds is 269 pm.

Fig. 4.7 The structures of the anions $[HFe_2(CO)_8]^-$ and $[HFe_3(CO)_{11}]^-$ determined as their $[(Ph_3P)_2N]^+$ salts.

The anion $[HRu_3(CO)_{11}]^-$ is prepared from $Ru_3(CO)_{12}$ (Eqn 4.4) and is structurally similar to its iron counterpart. At low temperature, protonation of $[HRu_3(CO)_{11}]^-$ with CF_3SO_3H gives $H_2Ru_3(CO)_{11}$, but this decomposes to $Ru_3(CO)_{12}$ at room temperature. Initial protonation occurs at the oxygen atom of the bridging carbonyl ligand as shown in Fig. 4.8.

$$Ru_3(CO)_{12} \xrightarrow[\text{2. } [Et_4N]Cl]{\text{1. Na}[BH_4], \text{ THF}} [Et_4N][HRu_3(CO)_{11}]$$

Eqn 4.4

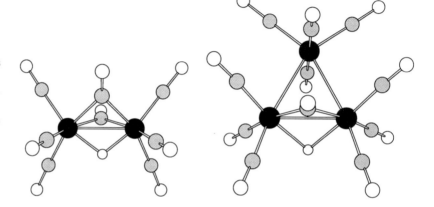

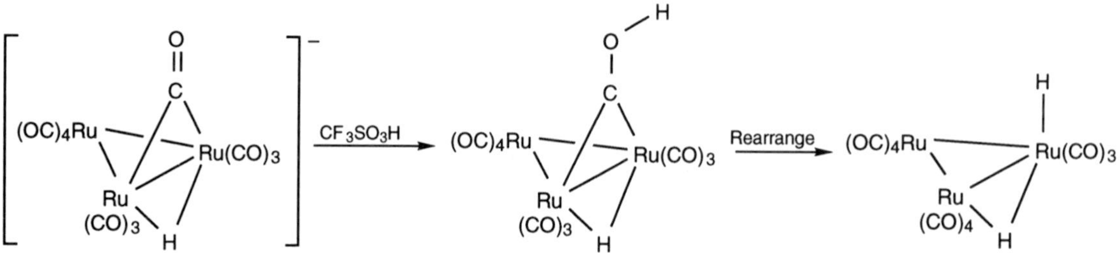

Fig. 4.8 The conversion of $[HRu_3(CO)_{11}]^-$ to $H_2Ru_3(CO)_{11}$; the reaction must be carried out at low temperature.

The reaction of $Ru_3(CO)_{12}$ and H_2 in boiling octane yields $H_4Ru_4(CO)_{12}$; it and the related cluster $H_2Ru_4(CO)_{13}$ are often found as by-products in reactions involving $Ru_3(CO)_{12}$ (or other ruthenium clusters) and hydrogen-containing reagents. Both $H_4Ru_4(CO)_{12}$ and $H_2Ru_4(CO)_{13}$ are 60-electron tetrahedral clusters (Fig. 4.9), and $H_2Ru_4(CO)_{13}$ can be converted to $H_4Ru_4(CO)_{12}$ by reaction with H_2 (Eqn 4.5).

$$H_2Ru_4(CO)_{13} + H_2 \xrightarrow{343\ K} H_4Ru_4(CO)_{12} + CO \qquad \textbf{Eqn 4.5}$$

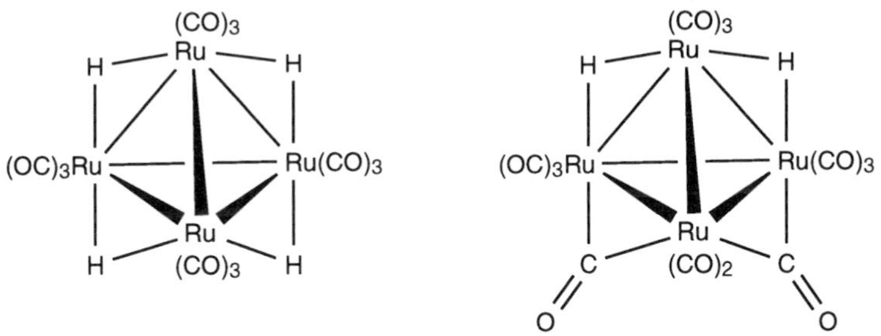

Fig. 4.9 The structures of $H_4Ru_4(CO)_{12}$ and $H_2Ru_4(CO)_{13}$.

In contrast to the triiron and triruthenium hydrido clusters described above, osmium forms the *unsaturated* compound $H_2Os_3(CO)_{10}$ — it has only 46 cluster bonding electrons despite adopting a triangular Os_3-core. This air-stable, purple compound can be prepared from $Os_3(CO)_{12}$ by reaction with H_2 (1 atm) at 398 K, but competitive degradation of the cluster can also occur (Eqn 4.6). The structure of $H_2Os_3(CO)_{10}$ is usually represented as shown in Fig. 4.10 and the Os=Os double bond emphasizes the unsaturation. Crystallographic data confirm that the hydride-bridged Os–Os bond is particularly short (268 pm). The addition of two-electrons to form a 48-electron cluster is observed when $H_2Os_3(CO)_{10}$ reacts with CO (Eqn 4.7), although further reaction with CO displaces the two hydride ligands as H_2 to give $Os_3(CO)_{12}$.

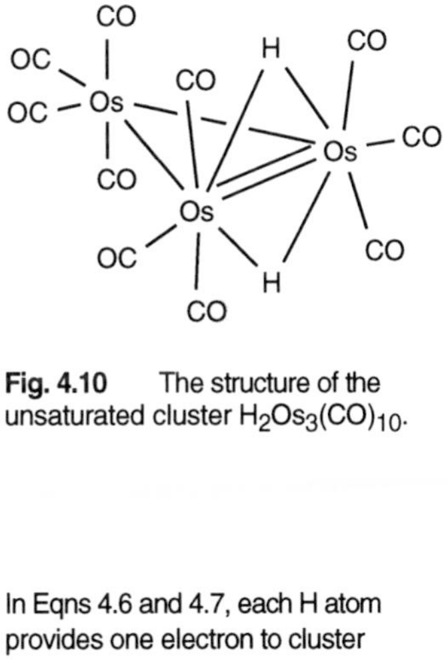

Fig. 4.10 The structure of the unsaturated cluster $H_2Os_3(CO)_{10}$.

$$Os_3(CO)_{12} \xrightarrow{H_2} H_2Os_3(CO)_{11} + H_2Os_2(CO)_8 + H_2Os(CO)_4 \qquad \textbf{Eqn 4.6}$$

$$H_2Os_3(CO)_{10} + CO \rightarrow H_2Os_3(CO)_{11} \qquad \textbf{Eqn 4.7}$$

In Eqns 4.6 and 4.7, each H atom provides one electron to cluster bonding, and the CO ligand provides two. Thus, the 2H-for-CO or CO-for-2H exchange occurs without a change to the overall number of electrons available for cluster bonding.

The structure of $H_2Os_3(CO)_{11}$ is analogous to that of $H_2Ru_3(CO)_{11}$ (Fig. 4.8), and the terminal and bridging hydride ligands undergo exchange on the ^1H NMR spectroscopic timescale. The conjugate base $[HOs_3(CO)_{11}]^-$ is formed as one of several products when $Os_3(CO)_{12}$ is treated with sodium tetrahydroborate (Eqn 4.8), and can be protonated under anhydrous conditions to give $H_2Os_3(CO)_{11}$.

$$Os_3(CO)_{12} \xrightarrow[THF]{Na[BH_4]} [HOs_3(CO)_{11}]^- + [H_2Os_4(CO)_{12}]^{2-} + [H_3Os_4(CO)_{12}]^-$$

$$\text{major product} \qquad\qquad \textbf{Eqn 4.8}$$

The parent cluster of the tetraosmium anions in reaction 4.8 is yellow $H_4Os_4(CO)_{12}$, and this is produced by the *prolonged* reaction of $Os_3(CO)_{12}$ with H_2 (1 atm), or using higher pressures of H_2. It is isostructural with $H_4Ru_4(CO)_{12}$ (Fig. 4.9). Osmium also forms the dihydride $H_2Os_4(CO)_{13}$.

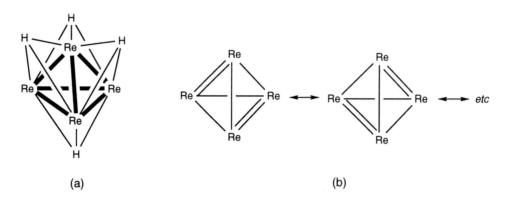

(a)   (b)

Fig. 4.11 (a) The H_4Re_4-core structure of $H_4Re_4(CO)_{12}$; each Re atom bears three terminal CO ligands. (b) The unsaturation of the cluster can be represented in valence bond terms, taking into account the fact that all the Re–Re bonds are of equal length.

Although in both $H_4Ru_4(CO)_{12}$ and $H_4Os_4(CO)_{12}$ the four hydride ligands are edge-bridging, in the rhenium cluster $H_4Re_4(CO)_{12}$, each hydride adopts a face capping mode (Fig. 4.11a). This may be rationalized in terms of the different electron counts in the $[M_4(CO)_{12}]^{4-}$ (M = Ru or Os) and $[Re_4(CO)_{12}]^{4-}$ cores. Whereas the group 8 cluster has the expected number of 60 electrons available, the electron count for the rhenium cluster is only 56. The cluster is therefore unsaturated as the resonance structures in Fig. 4.11b represent. Overall, the $[Re_4(CO)_{12}]^{4-}$ cluster core will have regions of electron density concentrated within the Re_3-faces rather than along the metal–metal edges as is the case for $[M_4(CO)_{12}]^{4-}$ (M = Ru or Os) — protonation of the former gives face-capping H ligands, whilst the latter is protonated to give edge bridging hydrides. The red compound $H_4Re_4(CO)_{12}$ may be prepared by reacting $Re_2(CO)_{10}$ with dihydrogen in a hydrocarbon solvent (> 373 K).

The reaction of $Ru_3(CO)_{12}$ with aqueous KOH in THF solution results in cluster expansion to the octahedral anion $[Ru_6(CO)_{18}]^{2-}$, protonation of which gives the monoanion or the neutral dihydrido-cluster depending on conditions (Eqn 4.9).

$$[Ru_6(CO)_{18}]^{2-} \xrightarrow[\text{in THF}]{\text{conc. } H_2SO_4} [HRu_6(CO)_{18}]^{-} \xrightarrow[\text{in CH}_2Cl_2]{\text{conc. } H_2SO_4} H_2Ru_6(CO)_{18}$$

Eqn 4.9

In $H_2Ru_6(CO)_{18}$, the hydride ligands cap opposite faces of the octahedral metal cage, but in the monoanion, the single H atom resides inside the Ru_6-cavity (Fig. 4.12). This interstitial mode has been confirmed by the results of a neutron diffraction study. As noted in Section 4.1, the interstitial hydride gives rise to an unusual chemical shift in the ^1H NMR spectrum of $[HRu_6(CO)_{18}]^-$ (δ +16.4).   On conversion from $[HRu_6(CO)_{18}]^-$ to $H_2Ru_6(CO)_{18}$, the H ligand already present in the cluster must migrate from the inside to the outside of the metal skeleton.

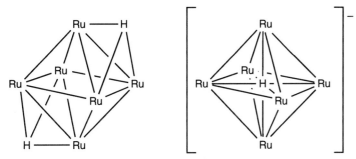

Fig. 4.12 The core structures of $H_2Ru_6(CO)_{18}$ and $[HRu_6(CO)_{18}]^-$ showing the positions of the hydride ligands. In each cluster, Ru represents an $Ru(CO)_3$ unit in which the ligands are terminally bonded.

Transition metal clusters containing an interstitial hydride are still few in number, and the mode of attachment is not considered 'normal' as would be μ- or $μ_3$ ligands. Two further examples are $[HRu_7(CO)_{20}]^-$ and $[HCo_6(CO)_{15}]^-$. The ruthenium cluster has a monocapped octahedral arrangement of metal atoms, with all the CO ligands terminally bound (Fig. 4.13). The anion $[HCo_6(CO)_{15}]^-$ is prepared by protonating $[Co_6(CO)_{15}]^{2-}$ at low temperature, but when $[(Ph_3P)_2N][HCo_6(CO)_{15}]$ is dissolved in water or methanol, the monoanion is unstable with respect to loss of the proton. Again, the migration of the proton through the metal skeleton appears to be quite facile.

4.3 Reactivity

The presence of a hydride ligand in a dimer or cluster is an important means of activating the compound. The simplest reaction is that of deprotonation, and this generates anions that are potential 'building-blocks' e.g. for cluster expansion reactions. Examples are given in Chapter 7. Other reaction types include the elimination of molecular dihydrogen, and the transfer of H atoms to unsaturated organic units (e.g. alkene to alkane transformation). Both these are illustrated in Chapter 6.

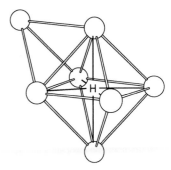

Fig. 4.13 The core structure and interstitial site of the hydride ligand in $[HRu_7(CO)_{20}]^-$; the structure was determined for the $[(Ph_3P)_2N]^+$ salt.

5 Terminal phosphine and bridging phosphido ligands

The substitution of one or more carbonyl groups for other ligands is commonly encountered, and large numbers of phosphine and phosphite derivatives of metal carbonyl dimers and clusters are known. A tertiary monodentate *phosphine* has the general formula PR_3, whilst a *phosphite* is of the form $P(OR)_3$. Each is a 2-electron donor and can replace a CO ligand without affecting the electron count at the metal centre. Related compounds containing arsine (AsR_3) and stibine (SbR_3) ligands may also be prepared by replacing terminal carbonyl groups.

Organophosphines:
Primary phosphine = PRH_2
Secondary phosphine = PR_2H
Tertiary phosphine = PR_3

5.1 Monodentate ligands

Substitution reactions of dimers $M_2(CO)_{10}$ containing group 7 metals are exemplified by considering $Mn_2(CO)_{10}$. This reacts with a range of phosphines, L, to give mono-substituted products $Mn_2(CO)_9L$, and there are two distinct sites into which the ligand can enter; these are defined in Fig. 5.1.

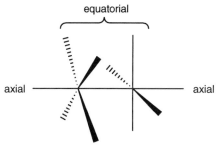

Fig. 5.1 Axial and equatorial sites in $Mn_2(CO)_{10}$ which has a staggered conformation. The axial sites are *trans* to the Mn–Mn bond, and the equatorial sites are *cis* to the Mn–Mn bond. See also Fig. 1.2.

Reactions of $Mn_2(CO)_{10}$ and phosphines, (for example, under photolytic conditions), to give both axially and equatorially substituted products are known, and two examples are $Mn_2(CO)_9(PPh_2H)$ (equatorial substitution) and $Mn_2(CO)_9(PMe_3)$ (axial substitution), the structures of which are shown in Fig. 5.2. Note that in both, the product retains a staggered conformation. With monodentate phosphines or phosphites, the introduction of a second ligand usually occurs at the second metal centre — that is, the product has the form $\{Mn(CO)_4L\}_2$. The substitution pattern tends to be axial-axial or equatorial-equatorial, and Fig. 5.3 illustrates the structure of the trimethylphosphite derivative $Mn_2(CO)_8\{P(OMe)_3\}_2$ — again, a staggered conformation is adopted. Both the steric and electronic requirements of the ligands play important roles in determining the site preferences and solid state structures of such derivatives.

The steric requirements of a phosphine or phosphite group can be assessed using the *Tolman cone angle* for the ligand. See the companion Primer *Organometallics 1: Complexes with Transition Metal–Carbon σ-Bonds* M. Bochmann, 1994.

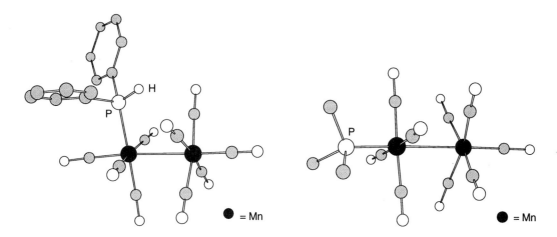

Fig. 5.2 The crystallographically determined structures of $Mn_2(CO)_8(PPh_2H)$ and $Mn_2(CO)_9(PMe_3)$, illustrating equatorial and axial substitution respectively. Hydrogen atoms, with the exception of the P-bound atom, are omitted.

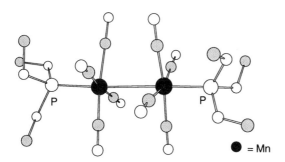

Fig. 5.3 The solid state structure of $Mn_2(CO)_8\{P(OMe)_3\}_2$; hydrogen atoms are omitted.

The mechanism by which CO substitution in $Mn_2(CO)_{10}$ occurs may be a dissociative process involving loss of CO and the formation of an unsaturated $\{Mn_2(CO)_9\}$ species (Eqn 5.1), or may take place by homolytic Mn–Mn bond cleavage (Eqn 5.2).

$$Mn_2(CO)_{10} \xrightarrow{-CO} \{Mn_2(CO)_9\} \xrightarrow{L} Mn_2(CO)_9L \qquad \textbf{Eqn 5.1}$$

$$\left.\begin{array}{l} Mn_2(CO)_{10} \rightarrow 2\,Mn(CO)_5^\bullet \\[4pt] Mn(CO)_5^\bullet + L \rightarrow Mn(CO)_4L^\bullet + CO \\[4pt] Mn(CO)_5^\bullet + Mn(CO)_4L^\bullet \rightarrow Mn_2(CO)_9L \end{array}\right\} \qquad \textbf{Eqn 5.2}$$

Steric crowding can be reduced by using a didentate ligand in place of two monodentate ligands, and this aspect of phosphine coordination is discussed further in Section 5.2.

The P–H bond of a terminally bound secondary phosphine ligand PR_2H has the possibility of undergoing intramolecular oxidative addition at an adjacent metal centre, and this leads to the formation of μ-PR_2 (*phosphido*)

Bridging
phosphido group = μ-PR_2

ligands. When $Mn_2(CO)_9(PPh_2H)$ is heated, the terminal PPh_2H group undergoes P–H bond cleavage as new Mn–P and Mn–H bonds are formed. In order that each Mn atom remains an 18-electron centre, one molecule of CO is ejected (Fig. 5.4). The product is $Mn_2(CO)_8(\mu\text{-}H)(\mu\text{-}PPh_2)$ and its solid state structure is shown in Fig. 5.5a. This type of reaction has been used quite widely to generate edge bridging phosphido groups in both dimers and clusters, and further examples are given in Eqns 5.3 and 5.4; in both products, two metal–metal bonds are each bridged by both a hydrido and phosphido ligand (Fig. 5.5b). By using a *primary* phosphine ligand, double P–H oxidative addition can occur on a trimetallic framework to give a $\mu_3\text{-}PR$ group.

μ_3-PR group: see Section 7.1.

$$Ru_3(CO)_{10}(PPh_2H)_2 \xrightarrow{\Delta} Ru_3(CO)_8(\mu\text{-}H)_2(\mu\text{-}PPh_2)_2 + 2CO \qquad \text{Eqn 5.3}$$

$$[Fe_4(CO)_{13}]^{2-} \xrightarrow[\text{in the presence of PPh}_2\text{H}]{CF_3CO_2H \text{ in THF}} Fe_3(CO)_8(\mu\text{-}H)_2(\mu\text{-}PPh_2)_2 +$$
$$Fe(CO)_4(PPh_2H) + CO \qquad \text{Eqn 5.4}$$

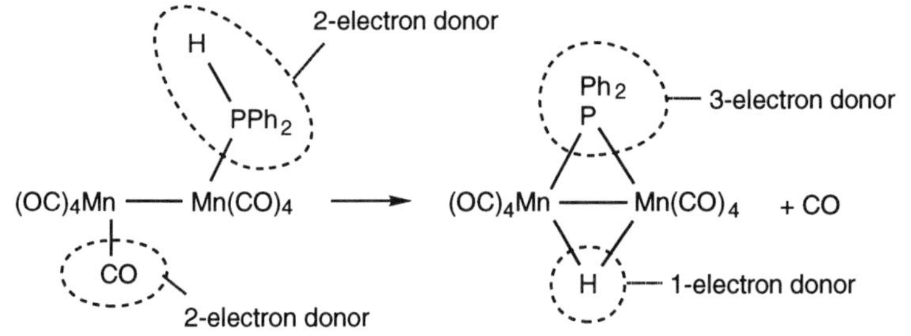

Fig. 5.4 In $Mn_2(CO)_9(PPh_2H)$, each ligand is a 2-electron donor and each Mn centre obeys the 18-electron rule. Upon oxidative addition of the P–H bond, the newly formed PPh_2 unit is a 3-electron donor and the H atom can contribute one electron. In order that each Mn atom remains an 18-electron centre, a molecule of CO must be lost.

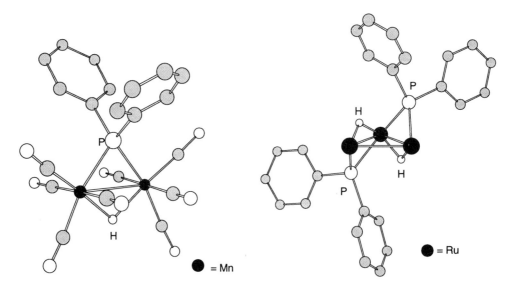

Fig. 5.5 (a) The crystallographically determined structure of $Mn_2(CO)_8(\mu\text{-}H)(\mu\text{-}PPh_2)$, and (b) the solid state structure of $Ru_3(CO)_8(\mu\text{-}H)_2(\mu\text{-}PPh_2)_2$, with CO ligands omitted — all are terminally bonded. The H atoms are omitted from the phenyl rings in both structures.

The replacement of a carbonyl by a phosphine or phosphite ligand is often carried out by photolysing or heating the two reagents in solution. The photolysis of $Ru_3(CO)_{12}$ leads to the loss of a CO ligand and the formation of the coordinatively unsaturated $Ru_3(CO)_{11}$ which is able to react with a Lewis base such as PPh_3. However, if CO is still available, $Ru_3(CO)_{11}$ reacts faster with this Lewis base than with PPh_3 and there is competition between the formation of $Ru_3(CO)_{11}(PPh_3)$ and $Ru_3(CO)_{12}$. Alternatively, the carbonyl compound can be activated by using either sodium benzophenone ketyl, $Na^+[Ph_2CO^{\bullet-}]$ (Eqn 5.5) or by replacing one or more carbonyls with the more labile acetonitrile prior to reaction with the group 15 ligand (Eqn 5.6).

$$Ru_3(CO)_{12} \xrightarrow[\text{L}]{\text{sodium benzophenone ketyl}} Ru_3(CO)_{11}L + Ru_3(CO)_{10}L_2$$

Eqn 5.5

$$Os_3(CO)_{12} \xrightarrow[\text{MeCN}]{Me_3NO} \begin{cases} Os_3(CO)_{11}(NCMe) \xrightarrow{L} Os_3(CO)_{11}L + MeCN \\ Os_3(CO)_{10}(NCMe)_2 \xrightarrow{L} Os_3(CO)_{10}L_2 + 2MeCN \end{cases}$$

Eqn 5.6

Use of amine oxide: see margin note on page 18.

Multi-substitution in clusters by monodentate phosphine or phosphite ligands generally takes place at successive metal atoms. The first Lewis base tends to enter at the least sterically crowded site and when $Fe_3(CO)_{12}$ reacts with triphenylphosphine, two isomers of $Fe_3(CO)_{11}(PPh_3)$ are produced in which the PPh_3 ligand attacks one of the two distinct metal sites in $Fe_3(CO)_{12}$ (Fig. 2.10). In each case, substitution occurs at a site *trans* to an Fe–Fe bond. In the disubstituted compound $Fe_3(CO)_{10}\{P(OMe)_3\}_2$ (Fig. 5.6), the two preferred sites are exactly those observed in the two isomers of $Fe_3(CO)_{11}(PPh_3)$. The pattern continues in the trisubstituted cluster $Fe_3(CO)_9(PMe_2Ph)_3$ — the third ligand favours a site on the third iron atom that is *trans* to an Fe–Fe bond. Notice that in each of these derivatives, the orientations of the remaining CO ligands reflect those of the parent carbonyl cluster $Fe_3(CO)_{12}$. Similarly, when $Co_4(CO)_{12}$ reacts with PPh_3, the structure of $Co_4(CO)_{11}(PPh_3)$ (Fig. 5.7) retains the three bridging CO ligands present in the parent compound (Fig. 2.16). In $Co_4(CO)_{10}\{P(OEt)_3\}_2$, the position of the first phosphite ligand mimics that of PPh_3 in $Co_4(CO)_{11}(PPh_3)$, and the second ligand enters in a similar site but at a different Co atom (Fig. 5.7).

Substitution without reorganization of the residual carbonyl framework does not always occur and this is illustrated by derivatives of $Ir_4(CO)_{12}$. This cluster reacts with triphenylphosphine to give $Ir_4(CO)_{11}(PPh_3)$, $Ir_4(CO)_{10}(PPh_3)_2$, and $Ir_4(CO)_9(PPh_3)_3$. With a smaller phosphine such as PEt_3, a tetrasubstituted derivative may be isolated. The solid state structures of representative members of the family $Ir_4(CO)_{12-x}L_x$ ($x = 1–4$, $L = PR_3$ or $P(OR)_3$) show a pattern of three μ-CO ligands and five, six, seven or eight terminal carbonyl groups. This contrasts with the structure of $Ir_4(CO)_{12}$ in which all the CO ligands are terminally bound.

O⁻
|
C
Ph   Ph

Benzophenone ketyl

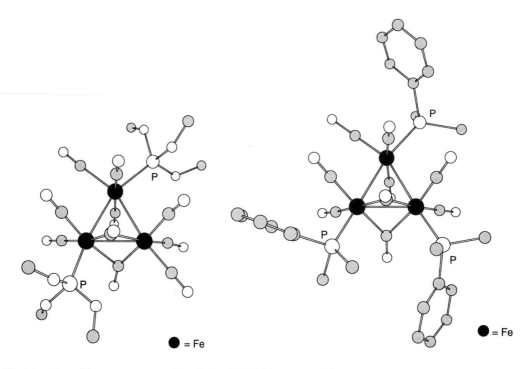

Fig. 5.6 The solid state structures of $Fe_3(CO)_{10}\{P(OMe)_3\}_2$ and $Fe_3(CO)_9(PMe_2Ph)_3$. Hydrogen atoms are omitted.

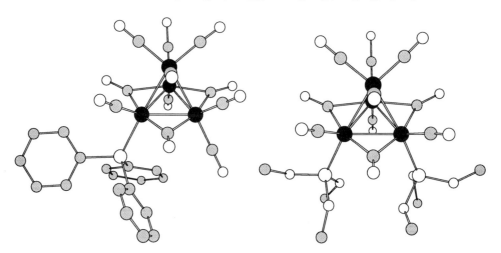

Fig. 5.7 The solid state structures of $Co_4(CO)_{11}(PPh_3)$ and $Co_4(CO)_{10}\{P(OEt)_3\}_2$. Hydrogen atoms are omitted.

5.2 Didentate ligands

We have seen that monodentate ligands tend to occupy sites remote from one another when more than one substituent is present in a dimer or cluster. The potential bonding modes of *polydentate* ligands are restricted by their inherent structure and several examples of didentate phosphine ligands are shown in Fig. 5.8.

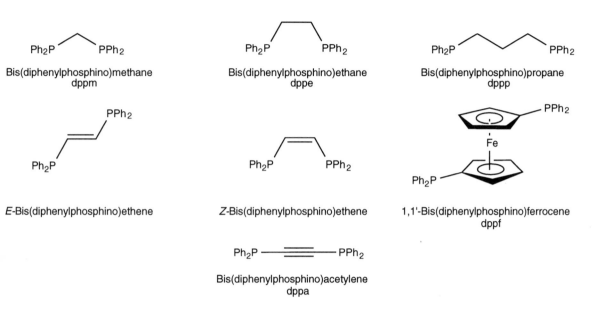

Ph₂P ⌄ PPh₂

Bis(diphenylphosphino)methane
dppm

Ph₂P ⌄ PPh₂

Bis(diphenylphosphino)ethane
dppe

Ph₂P ⌄ PPh₂

Bis(diphenylphosphino)propane
dppp

E-Bis(diphenylphosphino)ethene

Z-Bis(diphenylphosphino)ethene

1,1'-Bis(diphenylphosphino)ferrocene
dppf

Ph₂P ══ PPh₂

Bis(diphenylphosphino)acetylene
dppa

Fig. 5.8 Examples of didentate *P*-donor ligands with differing degrees of flexibility. Commonly used abbreviations for some of the ligands are given.

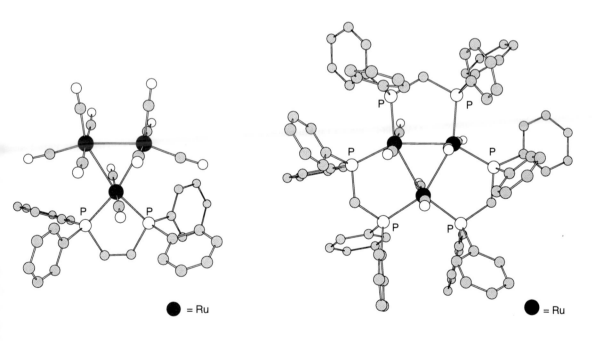

Fig. 5.9 The solid state structures of Ru₃(CO)₁₀L where L is *Z*-bis(diphenylphosphino)ethene, and Ru₃(CO)₆(μ-dppm)₃; H atoms are omitted. Notice that the 5-membered chelate ring in Ru₃(CO)₁₀L is planar, being fixed by the geometry of the phosphine ligand.

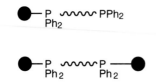

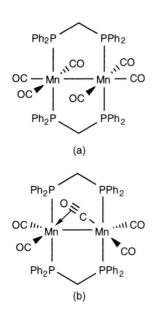

Fig. 5.10 Schematic representations of the pendant and linking modes of attachment of a didentate bis(diphenylphosphino)-ligand to metal centres.

Fig. 5.11 The structures of (a) $Mn_2(CO)_6(\mu\text{-dppm})_2$, and (b) $Mn_2(CO)_5(\mu\text{-dppm})_2$.

Some ligands are very flexible — for example, dppf is able to adjust its bite angle by the rotation of the two $\eta^5\text{-}C_5H_4PPh_2$ rings relative to one another; the rings can be staggered (as in the free ligand shown in Fig. 5.8) or eclipsed, or at an intermediate orientation. Ligands such as dppe, dppp and dppf are flexible enough to adopt different modes of attachment to a cluster — chelation to a single metal centre, bridging a metal–metal edge, or linking two metal units together. In contrast, the ligands derived from ethene and ethyne (acetylene) have rigid conformations. Z-Bis(diphenylphosphino)ethene is ideally suited to chelation at a single metal centre where it can form a planar five-membered chelate ring as in $Ru_3(CO)_{10}L$ (Fig. 5.9).

The rigidity of the ligands E-bis(diphenylphosphino)ethene and $Ph_2PC{\equiv}CPPh_2$ means that they will tend to adopt either a pendant mode (in which only one P atom is coordinated) or may link together two metal dimers or clusters by binding through one P-donor atom per metal unit (Fig. 5.10). These modes are readily distinguished by using ^{31}P NMR spectroscopy. For example, relative to 85% H_3PO_4 (a standard reference for ^{31}P NMR spectroscopy), the free dppf ligand exhibits one signal at δ –21.5. For a pendant dppf ligand attached to a ruthenium centre in a cluster, two ^{31}P NMR signals are observed at $\approx\delta$ –33 (coordinated phosphorus) and $\approx\delta$ –20 (non-coordinated phosphorus). When the dppf ligand is linking two ruthenium clusters together, one ^{31}P NMR resonance is observed at $\approx\delta$ –33.

Complexes of both metal dimers and clusters involving the dppm ligand are well exemplified. The carbon backbone of the ligand is too short for a chelating mode to be favourable, but its geometry is ideal to permit the dppm ligand to bridge between adjacent metal centres. The cluster $Ru_3(CO)_6(\mu\text{-dppm})_3$ (Fig. 5.9) illustrates attachment within the plane containing the Ru_3-triangle, a site preference that minimizes steric hindrance between the ligands. The reaction of $Mn_2(CO)_{10}$ with dppm leads to the formation of $Mn_2(CO)_6(\mu\text{-dppm})_2$ (Fig. 5.11a). An interesting feature of this dimer is that it readily loses CO (perhaps to relieve steric crowding) and in doing so, the electron count at one manganese centre is formally reduced to sixteen. However, one CO ligand 'leans across' and donates π-electron density to the otherwise coordinatively unsaturated metal atom (Fig. 5.11b). This is not a formal bridging mode, but rather a terminal CO ligand which additionally acts as a π-electron donor. The dppm ligands play an important role in supporting and stabilizing the dimetallic core, and further examples are seen in the family of diplatinum and dipalladium 'A-frame' complexes which includes $[Pt_2(CO)_2(\mu\text{-dppm})_2]^{2+}$, $Pt_2Cl_2(\mu\text{-dppm})_2$, $[Pt_2Cl_2(\mu\text{-NO})(\mu\text{-dppm})_2]^+$, $[Pt_2Cl_2(\mu\text{-H})(\mu\text{-dppm})_2]^+$, $Pd_2Cl_2(\mu\text{-dppm})_2$, $PdPtCl_2(\mu\text{-dppm})_2$, and $Pd_2Cl_2(\mu\text{-S})(\mu\text{-dppm})_2$.

6 Metal carbonyl dimers and clusters with organic ligands

6.1 Introduction: The cluster-surface analogy

So far, we have focused on the structures and reactivity of transition metal dimers and clusters with carbonyl ligands. In this section we look at some compounds containing π-bonded organic ligands, reactions between carbonyl clusters and alkenes and alkynes, and clusters containing alkylidyne and related ligands in which a CR group caps a metal triangle. Before discussing these compounds, we take an aside to consider why some of these molecular species may be of interest.

Two metal frameworks, the M_3-triangle and the M_4-butterfly, feature regularly in organometallic cluster chemistry, and one reason for this is their relationship to available sites on metal surfaces. Figure 6.1 illustrates the occurrences of a triangular M_3-site on a flat surface, and a butterfly M_4-site on a 'stepped' surface.

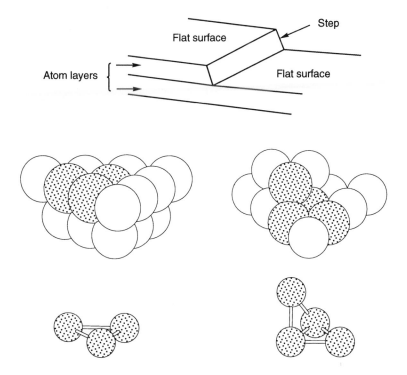

Fig. 6.1 In a bulk metal consisting of close-packed metal atoms, a layer of surface atoms may be flat, or may exhibit steps.
On a flat surface such as the (111) face of a face-centred cubic lattice, triangular M_3-sites are present. At a step, butterfly M_4-sites are available. The lower diagrams emphasize the M_3- and M_4- units, oriented as in the bulk metals drawn above. These same units are present in a range of metal carbonyl clusters — the CO ligands take the place of the surrounding metal atoms making the clusters models for the bulk metal sites.

The adsorption of small molecules such as H_2 and CO on to a metal surface can lead to dissociation, and this is driven by the formation of, for example, M–H and M–C bonds. The bonding interactions are not restricted to a single metal centre since metal atoms are in close proximity to each other. Adsorbed molecular fragments may play key roles during heterogeneous catalytic processes and there is evidence that the rhodium or platinum catalysed hydrogenation of ethene (Eqn 6.1) involves the formation and active participation of an ethylidyne unit (Fig. 6.2). It has been proposed that the CCH$_3$-unit faciliates the transfer of H atoms (formed from adsorbed dihydrogen) from the metal surface to molecular ethene.

$$CH_2{=}CH_2 + H_2 \xrightarrow{\text{Rh or Pt catalyst}} CH_3CH_3 \qquad \textbf{Eqn 6.1}$$

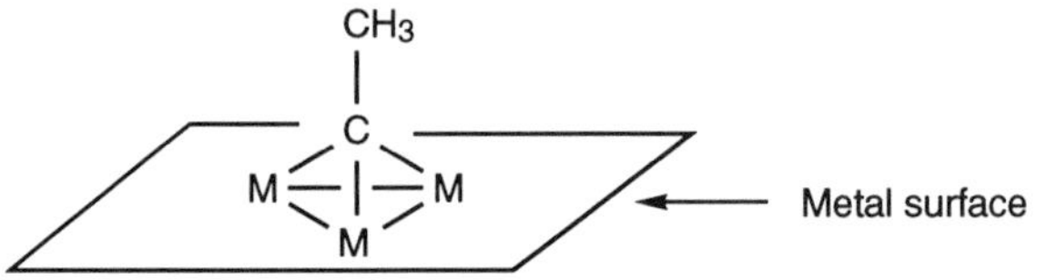

Fig. 6.2 An adsorbed ethylidyne fragment . This is proposed as a species present in the metal catalysed hydrogenation of ethene.

Attempts to use metal carbonyl clusters as models for heterogeneous catalysts have met with varying degrees of success, but one of the most well established results is that of Shriver and co-workers in which the proton induced reduction of a CO ligand to CH$_4$ models the conversion of carbon monoxide to methane on an iron or nickel surface. The sequence of cluster reactions is shown in Fig. 6.3.

See M.A. Drezdon *et al.* (1982), *Further reading* page 4.

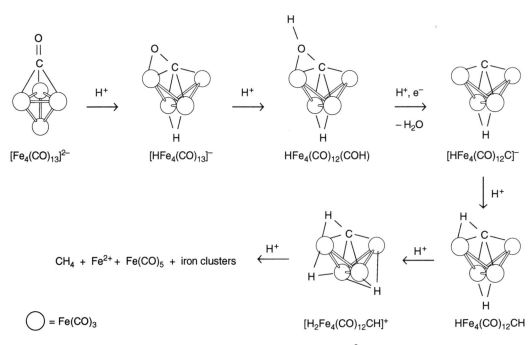

Fig. 6.3 The proton induced conversion of a μ_3-CO ligand in $[Fe_4(CO)_{13}]^{2-}$ to methane via a carbide cluster.

6.2 The cyclopentadienyl ligand

The cyclopentadienyl ligand Cp^- can be prepared from cyclopentadiene by reaction with an alkali metal or alkali metal hydride (Eqn 6.2).

C_5H_6 $+ NaH \xrightarrow{\text{THF}} H_2 + Na^+$ $[C_5H_5]^- = Cp^-$

Eqn 6.2

Alkali metal cyclopentadienide compounds (e.g. NaCp) react with a range of transition metal salts (e.g. $MnCl_2$, $FeCl_2$, $CoCl_2$) to give *sandwich complexes* of general formula Cp_2M. The preparation of nickelocene and ferrocene are shown in Eqns 6.3 and 6.4.

$$2C_5H_6 + NiCl_2 + 2KOH \rightarrow Cp_2Ni + 2KCl + 2H_2O$$

Eqn 6.3

$$2C_5H_6 + FeCl_2 + 2Et_2NH \rightarrow Cp_2Fe + 2Et_2NH_2Cl$$

Eqn 6.4

Note that although it is common to write simply Cp, it is more informative to include the hapticity of the ligand so that the formula for ferrocene becomes $(\eta^5\text{-}Cp)_2Fe$. The Cp^- ligand is flexible in its bonding, and the η^5-mode cannot be assumed in every instance.

We exemplify dimers involving cyclopentadienyl ligands by considering firstly complexes of iron and nickel, and then of chromium, molybdenum and tungsten. The reaction of C_6H_5 (generated *in situ*) with $Fe(CO)_5$ leads to the formation of the dimer $(\eta^5\text{-}Cp)_2Fe_2(CO)_4$ (Eqn 6.5), and the nickel dimer $(\eta^5\text{-}Cp)_2Ni_2(CO)_2$ is prepared according to reaction 6.6.

$$2C_5H_6 + 2Fe(CO)_5 \xrightarrow{\text{420 K}} (\eta^5\text{-}Cp)_2Fe_2(CO)_4 + 6CO + H_2$$

Eqn 6.5

$$Cp_2Ni + Ni(CO)_4 \rightarrow (\eta^5\text{-}Cp)_2Ni_2(CO)_2 + 2CO$$

Eqn 6.6

The structures of the two isomers of $(\eta^5\text{-}Cp)_2Fe_2(CO)_4$ are shown in Fig. 6.4, and Fig. 6.5 illustrates the solid state structure of $(\eta^5\text{-}Cp)_2Ni_2(CO)_2$. The $[\eta^5\text{-}Cp]^-$ ligand is a 6-electron donor, but it is often convenient to consider a *formally neutral* Cp ligand — a 5-electron donor. This method of counting means that no formal oxidation states need to be assigned to the metal centres. Thus in $(\eta^5\text{-}Cp)_2Fe_2(CO)_4$, each Fe centre obeys the 18-electron rule with five electrons from the Cp ligand, four in total from the CO ligands, eight in its valence shell and an Fe–Fe bond.

The structures of both the *cis*- and *trans*-isomers of $(\eta^5\text{-}Cp)_2Fe_2(CO)_4$ have been confirmed by X-ray crystallography. In solution, both isomers are present, and the terminal and bridging CO ligands undergo intramolecular exchange. Above 308 K, interconversion of the *cis*- and *trans*-forms occurs, and it is proposed that this involves the formation of an unbridged dimer.

Mononuclear cyclopentadienyl complexes and the bonding between a metal and an η^5-Cp ring are discussed in the companion Primer *Organometallics 2: Complexes with Transition Metal–Carbon π-Bonds* M. Bochmann, 1994.

Hapticity: see page 4.

The $[\eta^5\text{-}Cp]$ ligand is a 6-electron donor, but it is often convenient to consider a *formally neutral* Cp ligand — a 5-electron donor.

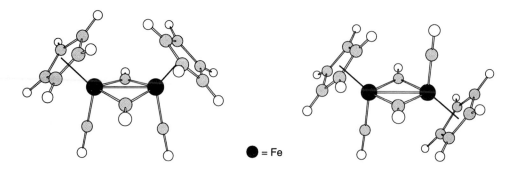

Fig. 6.4 The solid state structures of the *cis-* and *trans-*isomers of $(\eta^5\text{-Cp})_2\text{Fe}_2(\text{CO})_4$. The Fe–Fe bond length is 253 pm.

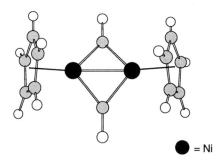

Fig. 6.5 The crystallographically confirmed structure of $(\eta^5\text{-Cp})_2\text{Ni}_2(\text{CO})_2$.
The Ni–Ni bond distance is 236 pm.

An aspect of the reactivity of $(\eta^5\text{-Cp})_2\text{Fe}_2(\text{CO})_4$ that is important is the ease of cleavage into monomeric units. Oxidation with halogens (Eqn 6.7) and reduction with sodium/mercury amalgam (Eqn 6.8) both generate mononuclear species which can be used in a range of syntheses to prepare organometallic derivatives containing $(\eta^5\text{-Cp})\text{Fe}$, $(\eta^5\text{-Cp})\text{Fe}(\text{CO})$ or $(\eta^5\text{-Cp})\text{Fe}(\text{CO})_2$-fragments, and these include heterometallic dinuclear compounds such as $(\eta^5\text{-Cp})\text{FeCo}(\text{CO})_6$ (Eqn 6.9).

$$(\eta^5\text{-Cp})_2\text{Fe}_2(\text{CO})_4 + \text{X}_2 \rightarrow 2(\eta^5\text{-Cp})\text{Fe}(\text{CO})_2\text{X} \qquad \text{X = Cl, Br or I} \qquad \textbf{Eqn 6.7}$$

$$(\eta^5\text{-Cp})_2\text{Fe}_2(\text{CO})_4 + 2\text{Na} \rightarrow 2\text{Na}[(\eta^5\text{-Cp})\text{Fe}(\text{CO})_2] \qquad \textbf{Eqn 6.8}$$

$$(\eta^5\text{-Cp})\text{Fe}(\text{CO})_2\text{Cl} + \text{Na}[\text{Co}(\text{CO})_4] \rightarrow (\eta^5\text{-Cp})\text{FeCo}(\text{CO})_6 + \text{NaCl} \qquad \textbf{Eqn 6.9}$$

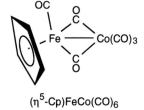

$(\eta^5\text{-Cp})\text{FeCo}(\text{CO})_6$

One or two *terminal* carbonyl ligands in $(\eta^5\text{-Cp})_2\text{Fe}_2(\text{CO})_4$ may be replaced by phosphines or phosphites. With isonitriles, substitution at both terminal and bridging sites occurs to give derivatives of the types shown in Fig. 6.6.

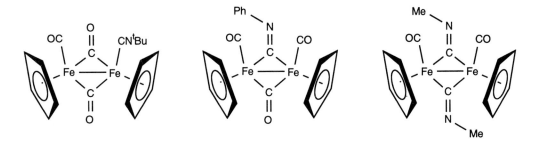

Fig. 6.6 Examples of the product types that can be obtained from the reactions of $(\eta^5\text{-Cp})_2\text{Fe}_2(\text{CO})_4$ with isocyanides.

One carbonyl bridge in $(\eta^5\text{-Cp})_2\text{Fe}_2(\text{CO})_4$ can be converted into a methylene unit as shown in Eqn 6.10 for the *trans*-isomer. The *cis* isomer is also present in solution, and interconverts with the *trans* compound. A hydride ion can be abstracted from the bridging CH_2-group to yield the cation $[(\eta^5\text{-Cp})_2\text{Fe}_2(\text{CO})_3(\mu\text{-CH})]^+$ (Eqn 6.11) which may be isolated as the $[\text{PF}_6]^-$ salt. The reaction of $(\eta^5\text{-Cp})_2\text{Fe}_2(\text{CO})_4$ with MeLi followed by protonation leads to formation of $(\eta^5\text{-Cp})_2\text{Fe}_2(\text{CO})_2(\mu\text{-CO})(\mu\text{-C}=\text{CH}_2)$ in both the *cis* and *trans* forms.

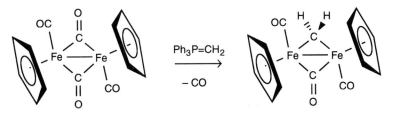

Eqn 6.10

Eqn 6.11

When $(\eta^5\text{-Cp})_2\text{Fe}_2(\text{CO})_4$ is heated (770 K), carbon monoxide is expelled, and the tetrairon cluster $(\eta^5\text{-Cp})_4\text{Fe}_4(\mu_3\text{-CO})_4$ is produced, but if the temperature is too high, ferrocene is the dominant product. The structure of $(\eta^5\text{-Cp})_4\text{Fe}_4(\mu_3\text{-CO})_4$ is shown in Fig. 6.7; the iron atoms define a tetrahedron and one CO ligand caps each face.

Fig. 6.7 The solid state structure of $(\eta^5\text{-Cp})_4Fe_4(\mu_3\text{-CO})_4$. Each CO ligand caps an Fe_3 face of the tetrahedral Fe_4-cluster core. The molecule is viewed looking along one of O–C bond vectors. Hydrogen atoms are omitted.

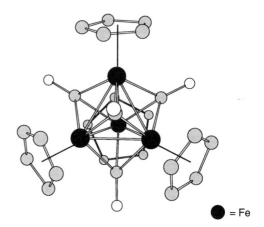

● = Fe

Electron counting for $(\eta^5\text{-Cp})_4Fe_4(\mu_3\text{-CO})_4$:

4 Fe =	4 × 8 = 32
4 CO =	4 × 2 = 8
4 η^5-Cp =	4 × 5 = 20
Total count =	60 electrons

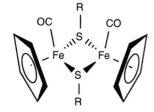

Fig. 6.8 The structure of $cis\text{-}(\eta^5\text{-Cp})_2Fe_2(CO)_2(\mu\text{-SR})_2$.

In some reactions of $(\eta^5\text{-Cp})_2Fe_2(CO)_4$, the Fe–Fe bond is cleaved but a dinuclear structure is retained with the dimetal core supported by bridging ligands. One example is the reaction with RSSR (e.g. R = Me, Ph) which gives $(\eta^5\text{-Cp})_2Fe_2(CO)_2(\mu\text{-SR})_2$ (Fig. 6.8). The *cis*-isomer is favoured over the *trans*-form, and crystallographic data show that the Fe⋯Fe separation is ≥ 300 pm, as opposed to the bonding distance of 253 pm in $(\eta^5\text{-Cp})_2Fe_2(CO)_4$. When $(\eta^5\text{-Cp})_2Fe_2(CO)_4$ reacts with elemental sulfur, the *cubane* cluster $(\eta^5\text{-Cp})_4Fe_4S_4$ is formed. (This is related to the cubane shown in Fig. 1.6). The Fe_4S_4-core of $(\eta^5\text{-Cp})_4Fe_4S_4$ is distorted (Fig. 6.9) with the iron atoms drawn together in a pairwise manner to give only two Fe–Fe bonding interactions (265 pm).

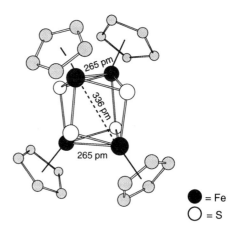

● = Fe
○ = S

Fig. 6.9 The structure of the compound $(\eta^5\text{-Cp})_4Fe_4S_4$; H atoms have been omitted.

The dimer $(\eta^5\text{-Cp})_2Ni_2(CO)_2$ is a purple diamagnetic compound, and the Ni–Ni distance (236 pm) is consistent with a bonding interaction. The absence of terminal CO ligands in $(\eta^5\text{-Cp})_2Ni_2(CO)_2$ means that CO-for-PR_3 substitution reactions of the type described for $(\eta^5\text{-Cp})_2Fe_2(CO)_4$ cannot occur, since *PR₃ ligands do not adopt bridging positions*. Instead $(\eta^5\text{-Cp})_2Ni_2(CO)_2$ reacts with 2-electron donor ligands with cleavage of the dimer (Eqn 6.12).

$$(\eta^5\text{-Cp})_2Ni_2(CO)_2 + PR_3 \rightarrow (\eta^5\text{-Cp})_2Ni + Ni(CO)_2(PR_3)_2 \qquad \textbf{Eqn 6.12}$$

The Ni–Ni bond in $(\eta^5\text{-Cp})_2Ni_2(CO)_2$ can be broken by chemical or electrochemical means. In a reaction similar to that of $(\eta^5\text{-Cp})_2Fe_2(CO)_4$, $(\eta^5\text{-Cp})_2Ni_2(CO)_2$ reacts with RSSR to yield $(\eta^5\text{-Cp})_2Ni_2(\mu\text{-SR})_2$ in which the metal dimer is supported by the bridging thiolate ligands. The reaction proceeds by cleavage of the dimer to give $(\eta^5\text{-Cp})Ni(CO)(SR)$ which undergoes dimerization and concomitant loss of CO. When P_2Ph_4 reacts with $(\eta^5\text{-Cp})_2Ni_2(CO)_2$, oxidative addition occurs and two phosphido bridges are formed (Eqn 6.13 and Fig. 6.10).

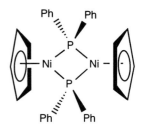

Fig. 6.10 The structure of the compound $(\eta^5\text{-Cp})_2Ni_2(\mu\text{-PPh}_2)_2$.

$$(\eta^5\text{-Cp})_2Ni_2(CO)_2 + Ph_2P\text{–}PPh_2 \rightarrow (\eta^5\text{-Cp})_2Ni_2(\mu\text{-PPh}_2)_2 \qquad \textbf{Eqn 6.13}$$

When $(\eta^5\text{-Cp})_2Ni_2(CO)_2$ is treated with sodium/potassium alloy, the paramagnetic, trinuclear cluster $(\eta^5\text{-Cp})_3Ni_3(\mu_3\text{-CO})_2$ is formed. The nickel atoms define a triangular array which is capped top and bottom by carbonyl ligands (Fig. 6.11). Reaction of the dimer with sulfur yields a structurally similar compound, $(\eta^5\text{-Cp})_3Ni_3(\mu_3\text{-S})_2$, in which capping sulfur atoms replace the $\mu_3\text{-CO}$ ligands. In $(\eta^5\text{-Cp})_3Ni_3(\mu_3\text{-CO})_2$, the Ni–Ni bonds are shorter (239 pm) than in $(\eta^5\text{-Cp})_3Ni_3(\mu_3\text{-S})_2$ (280 pm).

Iron and nickel are two places apart in the first row of the *d*-block — hence the difference between the number of carbonyls in $(\eta^5\text{-Cp})_2Fe_2(CO)_4$ and $(\eta^5\text{-Cp})_2Ni_2(CO)_2$ if the 18-electron rule is to be obeyed. If we move to chromium, (two places to the left of iron in the periodic table), the corresponding dimer is $(\eta^5\text{-Cp})_2Cr_2(CO)_6$ with three CO ligands per metal atom. The solid state structure of $(\eta^5\text{-Cp})_2Cr_2(CO)_6$ is shown in Fig. 6.12; the relatively long Cr–Cr single bond (328 pm) is *unsupported* by bridging CO ligands. The molybdenum and tungsten compounds $(\eta^5\text{-Cp})_2Mo_2(CO)_6$ and $(\eta^5\text{-Cp})_2W_2(CO)_6$ have analogous structures (Mo–Mo = 323 pm and W–W = 322 pm). In solution, the results of IR and variable temperature ^{1}H NMR spectroscopic studies indicate that there is an interconversion of the *anti* (as in Fig. 6.12) and *gauche* forms of each compound.

The dimer $(\eta^5\text{-Cp})_2Cr_2(CO)_6$ can be prepared by the oxidative coupling of $[(\eta^5\text{-Cp})Cr(CO)_3]^-$ (Eqns 6.14 and 6.15), and similar reactions can be used to give $(\eta^5\text{-Cp})_2M_2(CO)_6$ (M = Mo or W).

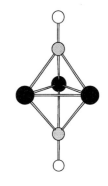

Fig. 6.11 The core structure of the compound $(\eta^5\text{-Cp})_3Ni_3(\mu_3\text{-CO})_2$; a cyclopentadienyl ligand is bonded to each nickel atom.

$$\text{NaCp} + Cr(CO)_6 \xrightarrow{\text{at reflux}} \text{Na}[(\eta^5\text{-Cp})Cr(CO)_3] + 3CO \qquad \textbf{Eqn 6.14}$$

$$2[(\eta^5\text{-Cp})Cr(CO)_3]^- \xrightarrow{\text{oxidative coupling}} (\eta^5\text{-Cp})_2Cr_2(CO)_6 \qquad \textbf{Eqn 6.15}$$

The Cr–Cr bond in $(\eta^5\text{-Cp})_2Cr_2(CO)_6$ is readily cleaved and dissociation occurs in solution to give paramagnetic $[(\eta^5\text{-Cp})Cr(CO)_3]^\bullet$ which is in equilibrium with the dimer. For the molybdenum and tungsten dimers, the stronger metal–metal bonds make such bond fission more difficult, although, for example, the $[(\eta^5\text{-Cp})Mo(CO)_3]^\bullet$ radical is formed when $(\eta^5\text{-Cp})_2Mo_2(CO)_6$ is photolysed.

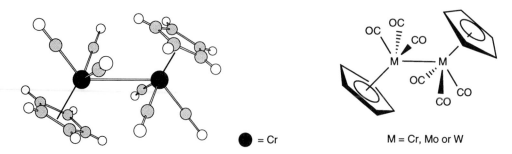

Fig. 6.12 The structure of $(\eta^5\text{-Cp})_2\text{Cr}_2(\text{CO})_6$, confirmed by X-ray crystallography, and a structural representation which emphasizes the presence of a metal–metal *single* bond in $(\eta^5\text{-Cp})_2\text{M}_2(\text{CO})_6$ (M = Cr, Mo or W) (compare with Fig. 6.13).

Carbonyl ligands in $(\eta^5\text{-Cp})_2\text{M}_2(\text{CO})_6$ (M = Cr, Mo or W) can be replaced by 2-electron donors including phosphine ligands (Eqn 6.16), but some reactions are complicated by the formation of ionic products of the type $[(\eta^5\text{-Cp})\text{Mo}(\text{CO})_2(\text{PR}_3)_2]^+[(\eta^5\text{-Cp})\text{Mo}(\text{CO})_3]^-$.

$$(\eta^5\text{-Cp})_2\text{Mo}_2(\text{CO})_6 \xrightarrow[-\text{CO}]{\text{PPh}_3, \ h\nu} (\eta^5\text{-Cp})_2\text{Mo}_2(\text{CO})_5(\text{PPh}_3) \ +$$

$$(\eta^5\text{-Cp})_2\text{Mo}_2(\text{CO})_4(\text{PPh}_3)_2 \qquad \textbf{Eqn 6.16}$$

Me

Me ⟨ ⟩ Me

Me Me

The pentamethylcyclopentadienyl ligand $[\eta^5\text{-C}_5\text{Me}_5]^-$ is often abbreviated to Cp*.

Loss of two CO ligands from $(\eta^5\text{-Cp})_2\text{Cr}_2(\text{CO})_6$ occurs when a toluene solution of the compound is heated, and the resultant unsaturation is manifested in the presence of a Cr≡Cr triple bond (Fig. 6.13). Structural data show this bond in $(\eta^5\text{-C}_5\text{Me}_5)_2\text{Cr}_2(\text{CO})_4$ to be significantly shorter (228 pm) than in $(\eta^5\text{-Cp})_2\text{Cr}_2(\text{CO})_6$ (328 pm). The Mo≡Mo bonded dimer $(\eta^5\text{-Cp})_2\text{Mo}_2(\text{CO})_4$ can be similarly prepared, and the photolysis of $(\eta^5\text{-Cp})_2\text{W}_2(\text{CO})_6$ results in the formation of $(\eta^5\text{-Cp})_2\text{W}_2(\text{CO})_4$. Notice that in Fig. 6.13, the CO ligands are oriented such that they lean towards the second metal atom; there is evidence that side-on C≡O-to-metal π-donation may occur.

When the unsaturated dimer $(\eta^5\text{-Cp})_2\text{M}_2(\text{CO})_4$ (M = Cr, Mo or W) is treated with CO, $(\eta^5\text{-Cp})_2\text{M}_2(\text{CO})_6$ is regenerated, and when PPh$_3$ is the Lewis base, *an addition rather than a substitution product* is obtained (Eqn 6.17). Diiodine adds across the unsaturated metal–metal bond in $(\eta^5\text{-Cp})_2\text{Mo}_2(\text{CO})_4$ to give $(\eta^5\text{-Cp})_2\text{Mo}_2\text{I}_2(\text{CO})_4$, and addition is also observed with alkynes as is illustrated in the next section.

Fig. 6.13 The solid state structure of $(\eta^5\text{-C}_5\text{Me}_5)_2\text{Cr}_2(\text{CO})_4$ (methyl groups are omitted), and a structural representation which emphasizes the presence of a metal–metal *triple* bond in $(\eta^5\text{-Cp})_2\text{M}_2(\text{CO})_4$ (M = Cr, Mo or W) (compare Fig. 6.12).

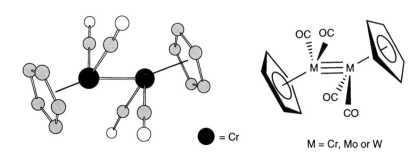

$$(\eta^5\text{-Cp})_2Mo_2(CO)_4 + 2PPh_3 \longrightarrow (\eta^5\text{-Cp})_2Mo_2(CO)_4(PPh_3)_2 \qquad \textbf{Eqn 6.17}$$

Clusters containing cyclopentadienyl ligands are numerous and can only be represented here by a few examples. Perhaps one of the most useful features to note when comparing these clusters to carbonyl species is that the introduction of an $M(\eta^5\text{-Cp})$ unit instead of an $M(CO)_3$ unit, means that the number of electrons available for bonding differs by one. Thus, whereas $Co(CO)_3$ has a total of 15 electrons available for cluster bonding, $Co(\eta^5\text{-Cp})$ can contribute only 14 electrons. This has structural ramifications — for a 60-electron tetrahedral cluster, four $Co(CO)_3$ units combine to give $Co_4(CO)_{12}$, but the combination of four $Co(\eta^5\text{-Cp})$ fragments provides only 56 electrons and additional hydrogen atoms make up the required electrons in $H_4(\eta^5\text{-Cp})_4Co_4$ (Eqn 6.18). Similarly, a tetrahedral iron cluster incorporating four $Fe(\eta^5\text{-Cp})$ units is eight electrons short of the 60-electron count, and the compound that is observed is $(\eta^5\text{-Cp})_4Fe_4(CO)_4$ (Fig. 6.7).

A CO ligand is a 2-electron donor.

η^5-Cp is a 5-electron donor – see margin note on page 55.

$$(\eta^5\text{-Cp})_2Co_2(NO)_2 \xrightarrow[\text{in presence of AlCl}_3]{\text{Li[AlH}_4]} (\eta^5\text{-Cp})_4Co_4(\mu_3\text{-H})_4 \qquad \textbf{Eqn 6.18}$$

Before leaving the cyclopentadienyl ligand, we should mention some related ligands with special features. Functionalizing the Cp⁻ ligand with a PR_2 group provides the ligand with an additional site through which is can bind to a metal atom. Equation 6.19 shows how an asymmetrical dimolybdenum complex can be formed. A related ligand is dppf shown in Fig. 5.8. By 'tying together' two cyclopentadienyl ligands, a dimer can be forced into a particular conformation in the solid state and show restricted molecular motion in solution. For example, reaction 6.20 gives a product that is related to $(\eta^5\text{-Cp})_2Fe_2(CO)_4$ but in which the two Cp-groups must adopt a *cis*-arrangement (compare Fig. 6.14 with Fig. 6.4).

$$\eqntext$$

Eqn 6.19

$$(\eta^5\text{-C}_5\text{H}_4)_2\text{SiMe}_2 + 2\text{Fe(CO)}_5 \xrightarrow[-\text{CO}]{\text{octane at reflux}} \{(\eta^5\text{-C}_5\text{H}_4)_2\text{SiMe}_2\}\text{Fe}_2(\text{CO})_4 \qquad \textbf{Eqn 6.20}$$

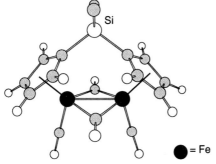

Fig. 6.14 The solid state structure of $\{(\eta^5\text{-C}_5\text{H}_4)_2\text{SiMe}_2\}\text{Fe}_2(\text{CO})_4$.

● = Fe

6.2   Reactions with alkynes

The reaction of the unsaturated dimer $(\eta^5\text{-Cp})_2\text{Mo}_2(\text{CO})_4$ with acetylene gives $(\eta^5\text{-Cp})_2\text{Mo}_2(\text{CO})_4(\text{HCCH})$ (Eqn 6.21) and illustrates the *addition* of an unsaturated organic molecule to an unsaturated metal–metal bond.

$$(\eta^5\text{-Cp})_2\text{Mo}_2(\text{CO})_4 + \text{RC}{\equiv}\text{CR} \rightarrow (\eta^5\text{-Cp})_2\text{Mo}_2(\text{CO})_4(\text{RCCR}) \qquad \textbf{Eqn 6.21}$$

e.g. R = H, Et, Ph

The structure of $(\eta^5\text{-Cp})_2\text{Mo}_2(\text{CO})_4(\text{HCCH})$ is shown in Fig. 6.15. The molybdenum atoms and the two carbon atoms derived from the alkyne define a distorted tetrahedral cluster core. The Mo–Mo and C–C distances (298 and 134 pm respectively) indicate bond orders that are lower than in the dimeric precursors. The molecular structure is made asymmetrical by the presence of one semi-bridging carbonyl ligand.

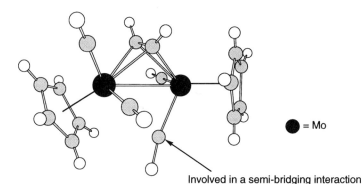

● = Mo

Involved in a semi-bridging interaction

Fig. 6.15 The solid state structure of $(\eta^5\text{-Cp})_2\text{Mo}_2(\text{CO})_4(\text{HCCH})$, the addition product of acetylene and $(\eta^5\text{-Cp})_2\text{Mo}_2(\text{CO})_4$.

Many carbonyl dimers possess metal–metal *single* bonds, and reactions between small unsaturated organic compounds and metal dimers take place with accompanying loss of carbonyl ligands or cleavage of the metal–metal bond. Reactions of $\text{Re}_2(\text{CO})_{10}$ with alkynes exemplify both these possibilities. When $\text{Re}_2(\text{CO})_{10}$ reacts with $\text{PhC}{\equiv}\text{CPh}$ at 460 K, products in which alkyne-coupling has occurred are formed. These include $\text{Re}_2(\text{CO})_7(\text{PhCCPh})_2$ and $\text{Re}_2(\text{CO})_6(\text{PhCCPh})_3$, proposed from spectroscopic data to have the structures shown in Fig. 6.16. The former structure-type has been confirmed by X-ray diffraction for the related compound $\text{Re}_2(\text{CO})_7\{\text{MeCC}(\text{NMe}_2)\text{C}(\text{NMe}_2)\text{CMe}\}$.

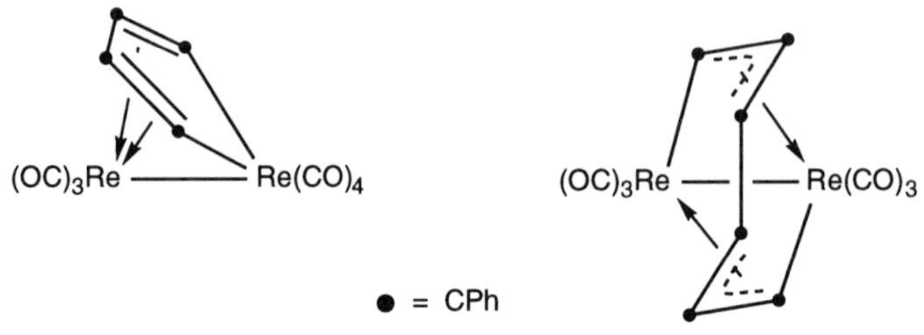

Fig. 6.16 The proposed structures of two products from the reaction between $\text{Re}_2(\text{CO})_{10}$ and $\text{PhC}{\equiv}\text{CPh}$. Alkyne coupling takes place, and the unsaturated organic ligand bonds to the dirhenium fragment by both σ- and π-interactions. Each Re atom is an 18-electron centre.

(OC)$_3$Re ——— Re(CO)$_4$

● = CPh

(OC)$_3$Re —|— Re(CO)$_3$

On the other hand, when $Re_2(CO)_9(NCMe)$ (formed from $Re_2(CO)_{10}$, Me_3NO and $MeCN$) reacts with $HC{\equiv}CCO_2Me$, the alkyne inserts into the Re–Re bond. In theory, a product of the type shown in Fig. 6.17a could form, but the presence of the CO_2Me substituent and the labile $MeCN$ ligand results in the formation of a chelate ring (Fig. 6.17b). Labelling studies using ^{13}C indicate that the insertion takes place without the precursor dissociating into mononuclear fragments.

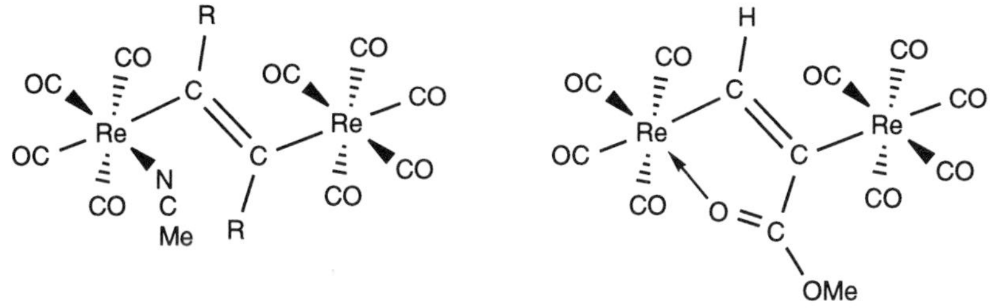

(a)   (b)

Fig. 6.17 (a) A possible product formed when $RC{\equiv}CR$ inserts into the Re–Re bond of $Re_2(CO)_9(NCMe)$. (b) The observed product of the reaction of $Re_2(CO)_9(NCMe)$ with $HC{\equiv}CCO_2Me$.

Reactions between $Co_2(CO)_8$ and $RC{\equiv}CR$ give clusters with Co_2C_2-cores (Eqn 6.22). A range of such compounds with differing R groups has been characterized, and the C–C distance is typically ≈ 135 pm. An alkyne ligand can be displaced by another alkyne which possesses substituents which are more electron-withdrawing, e.g. $F_3CC{\equiv}CCF_3$ replaces $PhC{\equiv}CPh$. An ordering of preferential attachment to the dicobalt framework is:

$$F_3CC{\equiv}CCF_3 > MeO_2CC{\equiv}CCO_2Me > PhC{\equiv}CPh > MeC{\equiv}CMe > HC{\equiv}CH.$$

The reaction of a diyne such as $PhC{\equiv}C–C{\equiv}CPh$ with $Co_2(CO)_8$ can result in attachment to two equivalents of metal dimer and the structure of $Co_2(CO)_6(PhCCCCPh)$ is shown in Fig. 6.18.

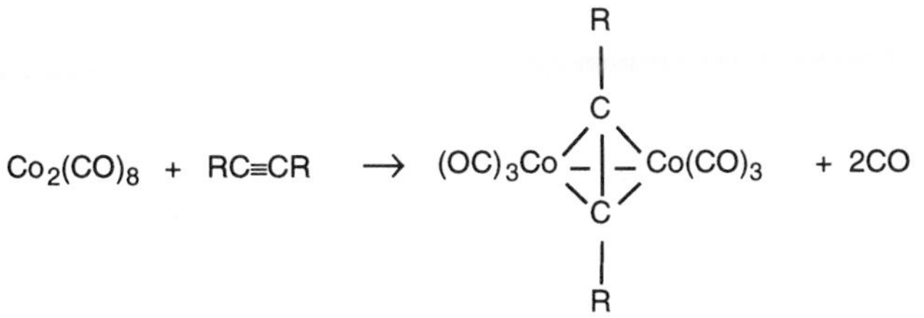

$$Co_2(CO)_8 \ + \ RC{\equiv}CR \quad \longrightarrow \quad (OC)_3Co\text{—}Co(CO)_3 \ + \ 2CO$$

Eqn 6.22

In $Co_2(CO)_6(RCCR)$, each cobalt atom is an 18-electron centre if the alkyne is counted as an overall 4-electron donor. Alternatively, by Wade's rules, each $Co(CO)_3$ and each CR unit provides 3 electrons making the tetrahedral Co_2C_2-core a 6-electron pair *nido*-cluster.

Further reaction between $Co_2(CO)_6(RCCR)$ and $RC{\equiv}CR$ (same R) leads to the expansion of the metal framework rather than the addition of another equivalent of alkyne (Eqn 6.23) and the product is a tetracobalt cluster in which the metal atoms define a butterfly structure; the Co_4C_2-core possesses a distorted octahedral geometry (Fig. 6.19).

$$Co_2(CO)_6(RCCR) + RC{\equiv}CR \ \longrightarrow \ Co_4(CO)_{10}(RCCR) + \text{other products} \qquad \textbf{Eqn 6.23}$$

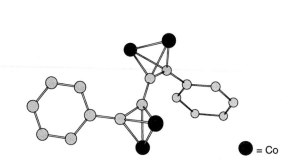

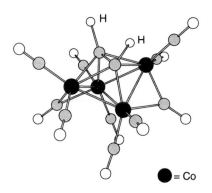

Fig. 6.18   The structure of the dicluster product {$Co_2(CO)_6$}$_2$(PhCCCCPh); H atoms and CO ligands have been omitted — all CO groups are terminally bonded.

Fig. 6.19   The structure of $Co_4(CO)_{10}$(HCCH); the crystallographic study was undertaken on the bis(triphenylphosphine) derivative.

The compound $Fe_2(CO)_9$ reacts with $^tBuC{\equiv}C^tBu$ (which has *bulky* substituents) to give $Fe_2(CO)_6(^tBuCC^tBu)$ and $Fe_2(CO)_4(^tBuCC^tBu)_2$ (Fig. 6.20). However, in the formation of $Fe_2(CO)_6(^tBuCC^tBu)$, the electron loss associated with the removal of three CO ligands does not balance the electron gain associated with each 4-electron donor alkyne ligand, and similarly for the formation of $Fe_2(CO)_4(^tBuCC^tBu)_2$. The products are formulated as having Fe=Fe double bonds, and this is supported by experimentally determined Fe–Fe bond distances of 222 pm. With less bulky organic substituents, the reactions of $Fe_2(CO)_9$ with alkynes can give 'flyover' complexes in which coupling of two equivalents of alkyne and a CO molecule takes place. An example of such a molecule is $Fe_2(CO)_6\{MeCC(Me)C(O)C(Me)CMe\}$, the structure of which is shown in Fig. 6.21. The organic ligand acts as a 6-electron donor, forming one C–Fe σ-bond and one C=C→Fe π-interaction per iron atom. Alternatively, coupling of two alkyne molecules may occur to give products such as $Fe_2(CO)_6\{HC=CHCH=CH\}$ (Fig. 6.22).

Fig. 6.20   The solid state structure of $Fe_2(CO)_6(^tBuCC^tBu)$ determined by X-ray crystallography; H atoms are omitted. Schematic representations of $Fe_2(CO)_6(^tBuCC^tBu)$ and $Fe_2(CO)_4(^tBuCC^tBu)_2$, showing the formal Fe=Fe double bond.

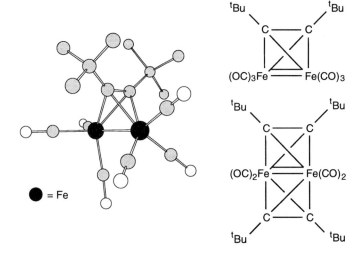

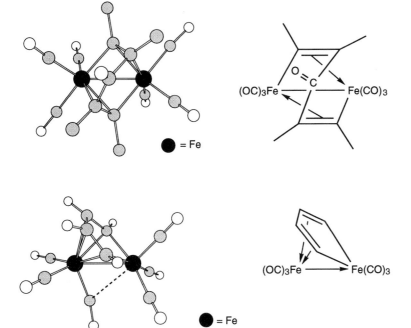

Fig. 6.21 The solid state structure of the compound Fe$_2$(CO)$_6${MeCC(Me)C(O)C(Me)CMe} with H atoms omitted; the schematic representation shows that the organic ligand is involved in one σ- and one π-interaction to each iron atom, making each metal an 18-electron centre.

Fig. 6.22 The solid state structure of Fe$_2$(CO)$_6${HC=CHCH=CH}; the schematic representation shows that one iron atom is involved in two Fe–C σ-interactions and the other iron atom in two π-interactions. An Fe→Fe coordinate bond is required to provide 18 electrons to each Fe centre. Compare this with the related compound in Fig. 6.16.

Alkynes of the type RC≡CR may react with Fe$_3$(CO)$_{12}$ to give products with a single RCCR ligand (Fig. 6.23), but in many cases, alkyne-coupling occurs to give more complex species. Transformations of organic fragments supported on metal frameworks are widely studied, and a relatively simple example is seen in the sequence of reactions beginning with the reaction of Fe$_3$(CO)$_9$(RCCR) with RC≡CR followed by pyrolysis. Initially, the second molecule of alkyne substitutes for a carbonyl ligand to give the violet compound Fe$_3$(CO)$_8$(RCCR)$_2$ (Fig. 6.24, R = Ph). When heated, this converts into a black isomer in which the two alkyne ligands are coupled and one Fe–Fe bond has been ruptured (Fig. 6.25, R = Ph). Reactions of Ru$_3$(CO)$_{12}$ or Os$_3$(CO)$_{12}$ with alkynes have been thoroughly studied, and as for iron, disubstituted alkynes may simply add with concomitant CO loss, or may also undergo coupling.

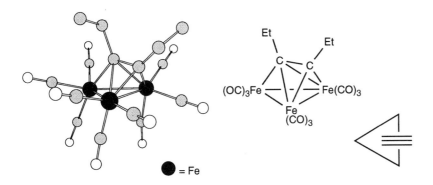

Fig. 6.23 The crystallographically determined structure of Fe$_3$(CO)$_9$(EtCCEt), and a schematic representation of the molecule. Note that the apparently 5-coordinate carbon atom emphasizes that the bonding is most easily treated in a delocalized manner. The lower right-hand schematic is used to illustrate why this mode of bonding is referred to as the 'π-perpendicular' mode — contrast with Fig. 6.24.

Fig. 6.24 The solid state structure of the violet isomer of $Fe_3(CO)_8(PhCCPh)_2$; only the *ipso*-carbon atom of each Ph ring is shown. The schematic representation of the structure shows that each organic ligand can be considered to form two σ- and one π-bond to the closed Fe_3-framework. This mode of attachment is called the 'π-parallel' mode — contrast with Fig. 6.23.

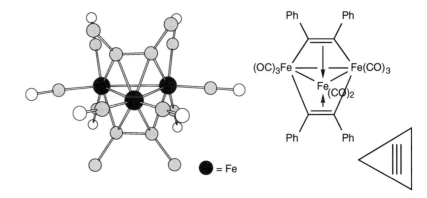

Fig. 6.25 The solid state structure of the black isomer of $Fe_3(CO)_8(PhCCPh)_2$; only the *ipso*-carbon atom of each Ph ring is shown. The alkyne ligands are coupled. The schematic representation of the structure shows that the C_4Ph_4 ligand can be considered to form both σ- and π-interactions with the open Fe_3-framework.

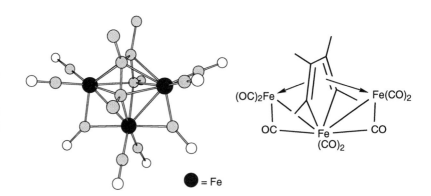

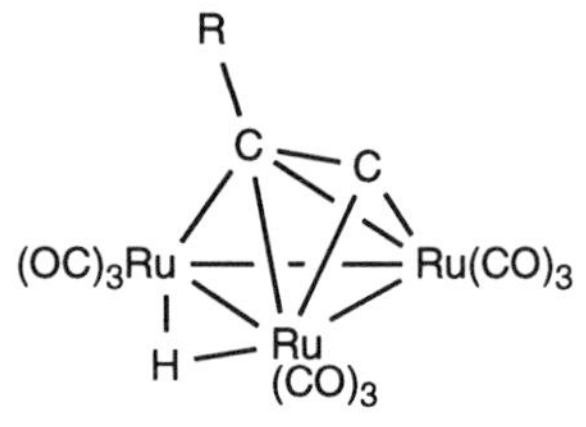

Terminal alkynes, RC≡CH, may give hydrido complexes when they react with a metal cluster, since the C–H bond can undergo oxidative addition to the metal framework. Thus, $Os_3(CO)_{12}$ reacts with PhC≡CH to yield $Os_3(CO)_{10}(PhCCH)$ and $Os_3(CO)_9(PhCCH)$, which convert to $HOs_3(CO)_9(CCPh)$ when heated, and $Ru_3(CO)_{12}$ reacts with RC≡CH to give $HRu_3(CO)_9(CCR)$ (Fig. 6.26). Oxidative addition of C–H bonds is also observed when alkenes react with clusters and examples are given in Section 6.3.

Higher nuclearity clusters have the possibility of undergoing skeletal rearrangements during reactions and can also facilitate fission of the alkyne into two alkylidyne, CR, units. This is observed in the reactions of RC≡CR with $Os_6(CO)_{16}(NCMe)_2$ (formed from $Os_6(CO)_{18}$ by reaction with Me_3NO and MeCN). The initial product is $Os_6(CO)_{16}(RCCR)$, and the alkyne ligand behaves as a 4-electron donor, binding to the cluster in a π-parallel mode. On heating this compound, the carbon-carbon bond is broken — the cluster framework is able to support the attachment of two CR groups, and each is a 3-electron donor. The additional two electrons could result in the loss of a CO ligand or in the cleavage of an Os–Os bond. The observed product confirms the latter (Fig. 6.27).

Fig. 6.26 The structure of $HRu_3(CO)_9(CCPh)$. Compare with the structure shown in Fig. 6.23.

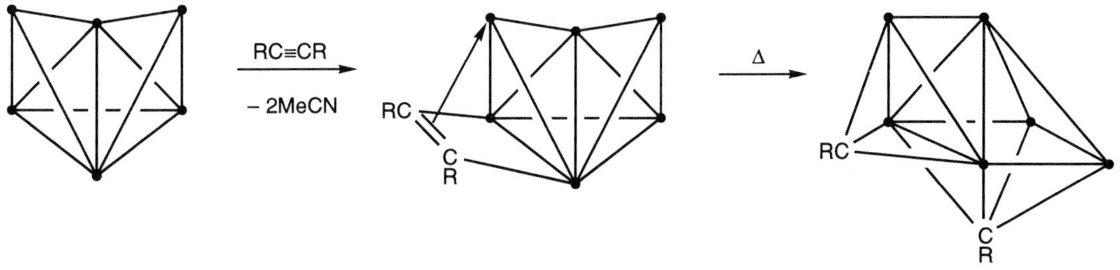

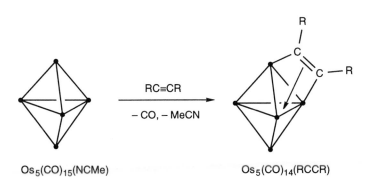

Os$_6$(CO)$_{16}$(NCMe)$_2$ Os$_6$(CO)$_{16}$(RCCR)   Os$_6$(CO)$_{16}$(CR)$_2$

Fig. 6.27 The reaction of Os$_6$(CO)$_{16}$(NCMe)$_2$ with RC≡CR followed by thermolysis causes carbon-carbon bond cleavage.

The fact that the alkyne initially 'sees' the hexaosmium cluster as no different than a metal triangle and that there is a structural similarity between the Fe$_3$C$_2$-unit in Fe$_3$(CO)$_8$(RCCR)$_2$ (Fig. 6.24) and the Os$_3$C$_2$-unit in Os$_6$(CO)$_{16}$(RCCR) (Fig. 6.27) is not a coincidence. The surfaces of many high-nuclearity clusters are composed of triangular faces and an analogy with the reactivity of M$_3$-clusters follows. A further example is the reaction between Os$_5$(CO)$_{15}$(MeCN) and alkynes, illustrated in Fig. 6.28.

Fig. 6.28 The reaction of Os$_6$(CO)$_{15}$(NCMe) with a disubstituted alkyne RC≡CR gives a product in which the alkyne bonds in a π-parallel mode.

Os$_5$(CO)$_{15}$(NCMe)   Os$_5$(CO)$_{14}$(RCCR)

6.3 Reactions with alkenes and dienes

An alkene RHC=CHR could function as a 2-electron donor and might be expected to replace a CO ligand in a metal dimer of cluster. However, many such reactions are more complex, being accompanied, for example, by C–H bond activation. In this section we look at representative examples of reactions of alkenes and dienes with dimers and clusters. Ligands that will be considered include cycloocta-1,5-diene (cod) and norbornadiene (nbd) the structures of which are shown in Fig. 6.29. Each diene can function as a 4-electron donor, replacing two CO ligands.

The thermolysis of Fe$_2$(CO)$_9$ with ethene results in the formation of the substituted *mononuclear* product Fe(CO)$_4$(η^2-C$_2$H$_4$) (Fig. 6.30). Similar results are observed when buta-1,3-diene is used.

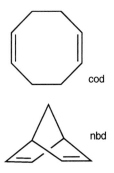

cod

nbd

Fig. 6.29 The dienes cycloocta-1,5-diene and norbornadiene. The ligands are referred to by the standard abbreviations cod and nbd.

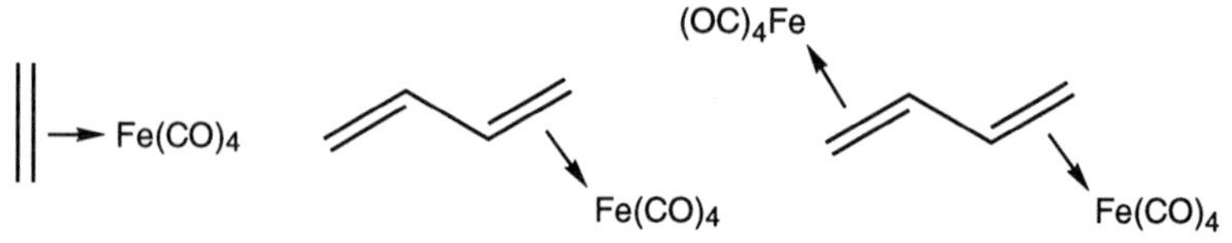

Fig. 6.30 Mononuclear products of the reactions of $Fe_2(CO)_9$ with ethene and buta-1,3-diene.

Trimetallic clusters containing group 8 metals show a variety of reaction types with acyclic and cyclic alkenes. When $Ru_3(CO)_{12}$ is photolysed with ethene, the product is $Ru(CO)_4(\eta^2\text{-}C_2H_4)$. The reaction is quantitative, and is a valuable source of the $Ru(CO)_4$-unit for the synthesis of heterometallic clusters. Equation 6.24 shows the reaction of $Ru(CO)_4(\eta^2\text{-}C_2H_4)$ with an iron dimer to give a mixed-metal cluster; further reaction leads to $Fe_2Ru_2(CO)_{11}(\mu_4\text{-Te})_2$, the structure of which is shown in Fig. 6.31. A further example is seen in the formation of the ditungsten-ruthenium compound in Eqn 6.25. Product **A** is formed in 80% yield, but if the reaction is carried out using $Ru_3(CO)_{12}$ instead of the alkene complex as the source of $Ru(CO)_x$, the yield is only 10%. By changing the reaction conditions, a higher nuclearity heterometallic cluster is obtained — this is product **B** in Eqn 6.25.

$$Ru(CO)_4(\eta^2\text{-}C_2H_4) + Fe_2(CO)_6(\mu\text{-Te})_2 \xrightarrow{-C_2H_4,\ -CO} Fe_2Ru(CO)_9(\mu_3\text{-Te})_2$$

Eqn 6.24

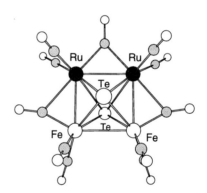

Fig. 6.31 The solid state structure of $Fe_2Ru_2(CO)_{11}(\mu_4\text{-Te})_2$.

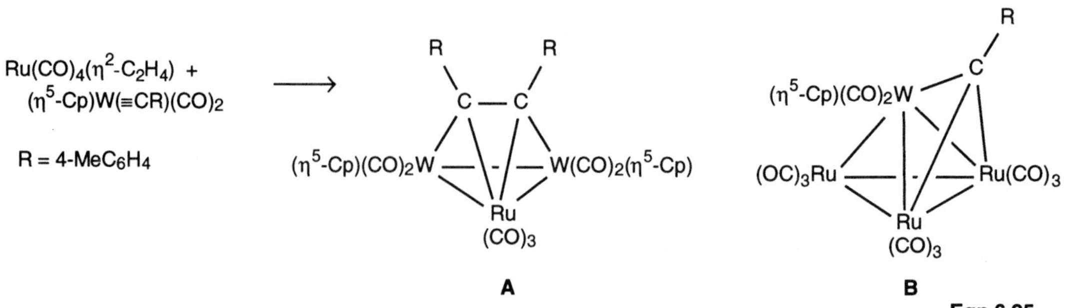

$$Ru(CO)_4(\eta^2\text{-}C_2H_4) + (\eta^5\text{-Cp})W(\equiv CR)(CO)_2 \longrightarrow$$

$R = 4\text{-MeC}_6H_4$

A

B

Eqn 6.25

Under mild conditions, it is possible to obtain the simple substitution product $Os_3(CO)_{11}(\eta^2\text{-}C_2H_4)$ from the reaction of $Os_3(CO)_{11}(NCMe)$ with ethene. At 398 K, alkenes $RHC=CH_2$ (R = H or alkyl) react with $Os_3(CO)_{12}$ to yield the isomeric clusters shown in Fig. 6.32. Similar reactions occur with $Ru_3(CO)_{12}$. Two C–H bonds undergo oxidative addition to the cluster and two M–H–M bridges are formed. Both isomers are fluxional in solution on the NMR spectroscopic timescale. When the reaction between $Os_3(CO)_{12}$ and ethene is carried out at 20 atm pressure and 430 K, the tetranuclear cluster $Os_4(CO)_{12}(HCCH)$ can be isolated along with $Os_4(CO)_{12}(HCCEt)$. These compounds have similar structures to that of $Co_4(CO)_{10}(RCCR)$ (Fig. 6.19), and the formation of $Os_4(CO)_{12}(HCCEt)$ indicates that alkene coupling has occurred.

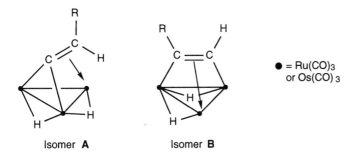

● = Ru(CO)$_3$
or Os(CO)$_3$

Isomer **A**   Isomer **B**

Fig. 6.32 The two isomers of $H_2M_3(CO)_9(HCCR)$ for M = Ru or Os. The iron analogue of isomer A is described in Section 6.4.

The addition of terminal alkenes $RHC=CH_2$ to the unsaturated cluster $H_2Os_3(CO)_{10}$ occurs with concomitant hydrogenation of a second mole of alkene (Eqn 6.26). The structure of the addition product is shown in Fig. 6.33. Under an atmosphere of H_2, $H_2Os_3(CO)_{10}$ behaves as a catalyst (albeit not a very active one) for alkene-to-alkane conversion. Various other transition metal clusters have been found to function as alkene hydrogenation catalysts, and these include $Os_3(CO)_{12}$ and $H_4Os_4(CO)_{12}$ (cyclohexene→cyclohexane), $H_3Os_4(CO)_{12}I$ (styrene→ethylbenzene), $H_4Cp_4Ni_4$ (hex-1-ene→hexane or pent-1-ene→pentane) and $H_3CpNiOs_3(CO)_9$ (3,3-dimethylbut-1-ene→3,3-dimethylbutane).

$$H_2Os_3(CO)_{10} + 2RHC=CH_2 \rightarrow HOs_3(CO)_{10}(CHCHR) + RCH_2CH_3 \quad \textbf{Eqn 6.26}$$

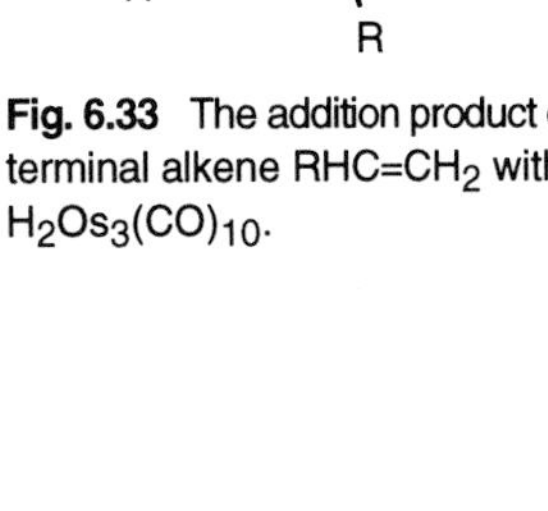

The reactions between $Ru_3(CO)_{12}$ and dienes lead to a range of products, and are generally non-specific. Much depends upon the relative positions of the C=C double bonds in the diene. For example, with hexa-2,4-diene, a mixture of isomers of $HRu_3(CO)_9(C_6H_9)$ (Fig. 6.34) is produced. Isomer **A** (containing an allene ligand) converts to **B** (with an allylic unit) upon being heated. With an excess of cyclohexa-1,3-diene, $Ru_3(CO)_{10}(NCMe)_2$ (prepared from $Ru_3(CO)_{12}$, MeCN and Me_3NO) reacts to give $HRu_3(CO)_9(C_6H_7)$ in which the organic ligand is bound to the metal framework as shown in Fig. 6.35; it acts as a 5-electron donor giving the metal cluster the 48 electron count required for a triangle. An analogous osmium compound can be formed from $H_2Os_3(CO)_{10}$ and cyclohexa-1,3-diene. Both $HRu_3(CO)_9(C_6H_7)$ and $HOs_3(CO)_9(C_6H_7)$ lose H^- when

(CO)$_4$
Os

(OC)$_3$Os————Os(CO)$_3$

C=C

R

Fig. 6.33 The addition product of a terminal alkene $RHC=CH_2$ with $H_2Os_3(CO)_{10}$.

treated with $[Ph_3C]^+$, and deprotonation of the resulting cations gives $M_3(CO)_9(C_6H_6)$ (M = Ru or Os, Eqn 6.27). The solid state structure of $Ru_3(CO)_9(C_6H_6)$ is shown in Fig. 6.36 and confirms the presence of a coordinated benzene ligand which exhibits a Kekulé-like distortion with alternating long (145 pm) and short (140 pm) C–C bonds.

$$HM_3(CO)_9(C_6H_7) \xrightarrow[-Ph_3CH]{[Ph_3C]^+} [HM_3(CO)_9(C_6H_6)]^+ \xrightarrow[-H^+]{DBU} M_3(CO)_9(C_6H_6)$$

M = Ru or Os

Eqn 6.27

DBU is 1,8-diazabicyclo[5.4.0]undec-7-ene.

Fig. 6.34 Two isomers of $HRu_3(CO)_9(C_6H_9)$ are formed when $Ru_3(CO)_{12}$ reacts with hexa-2,4-diene

Isomer **A**   Isomer **B**

Fig. 6.35 The solid state structure of $HRu_3(CO)_9(C_6H_7)$ with a schematic representation that provides a bonding description for the organic ligand-metal interactions. The ligand might also be represented in the form:

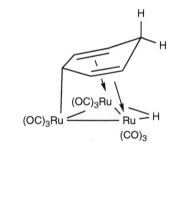

= Ru

Fig. 6.36 The solid state structure of $Ru_3(CO)_9(C_6H_6)$ with a schematic representation that emphasizes the triene-like nature of the ligand.

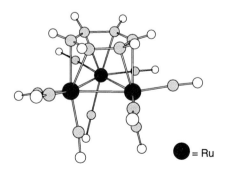

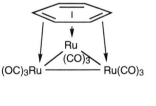

= Ru

The geometry of norbornadiene (Fig. 6.29) makes it ideally suited for coordination to two adjacent sites *at one metal centre*. When nbd reacts with $Co_2(CO)_8$ displacement of two or four CO ligands occurs — $Co_2(CO)_6(nbd)$ possesses an asymmetrical structure with one Co atom retaining 3 terminal CO ligands, whilst the structure of $Co_2(CO)_4(nbd)_2$ is symmetrical (Fig. 6.37).

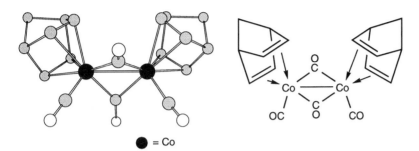

● = Co

Fig. 6.37 The solid state structure and a schematic representation of $Co_2(CO)_4(nbd)_2$.

The reaction between cycloocta-1,5-diene and $Ir_4(CO)_{12}$ in THF at reflux in the presence of Me_3NO (to assist in removal of CO) results in the substitution of two or four CO ligands and the formation of $Ir_4(CO)_{10}(cod)$ and $Ir_4(CO)_8(cod)_2$. Further reaction with cod, again in the presence of Me_3NO to assist in the removal of CO ligands, produces $Ir_4(CO)_6(cod)_3$. Each cod ligand substitutes at a different iridium centre. The reaction conditions are critical — if the solvent is chlorobenzene, the product is $Ir_4(CO)_5(cod)_2(C_8H_{10})$, the structure of which is shown in Fig. 6.38. Whilst two cod ligands coordinate in the expected manner, the addition of the third ligand is accompanied by the oxidative addition of two C–H bonds. Two carbon atoms are thus rendered an intimate part of the cluster core, and CO is eliminated; two H atoms are either released as H_2 or are transferred to a cod ligand resulting in alkene→alkane transformation (the outcome has not been confirmed).

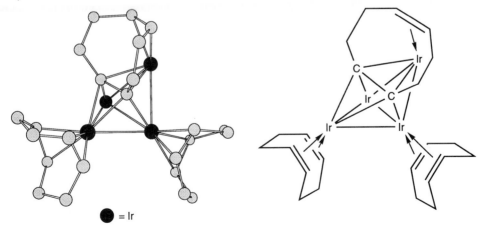

● = Ir

Fig. 6.38 The solid state structure of $Ir_4(CO)_5(cod)_2(C_8H_{10})$ and a schematic representation of the molecule; CO ligands have been omitted. Note that two C atoms of one organic ligand have inserted into the metal framework and the Ir_4C_2-unit resembles that in $Co_4(CO)_{10}(HCCH)$, shown in Fig. 6.19.

6.4 Alkylidyne and related ligands

Clusters containing a capping alkylidyne ligand, μ_3-CR (R = H or alkyl), are well established and in addition there is a range of related compounds in which the R group might be Cl, Br, CO_2H, OMe or aryl. Electron book-keeping in the clusters $Co_3(CO)_9CCl$, $H_3Fe_3(CO)_9CH$, $HFe_3(CO)_{10}CH$, $H_3Ru_3(CO)_9CMe$ and $H_3Os_3(CO)_9CCO_2H$ confirms the required 48-electron count for the metal triangle with the CR unit acting as a 3-electron ligand. The structures of two members of this group of clusters are shown in Fig. 6.39 — each possesses a tetrahedral M_3C-core.

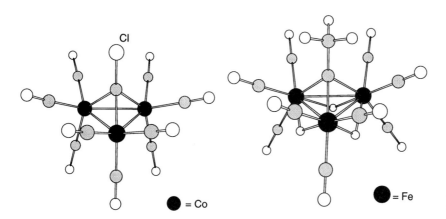

Cl

● = Co

● = Fe

Fig. 6.39 The solid state structures of $Co_3(CO)_9CCl$ and $H_3Fe_3(CO)_9CMe$.

Equation 6.28 shows a general strategy for the formation of clusters of type $Co_3(CO)_9CR$ (R = H, alkyl, halogen or aryl).

$$RCX_3 + Co_2(CO)_8 \rightarrow Co_3(CO)_9CR + CoX_2 \qquad \text{e.g. X = Cl} \qquad \textbf{Eqn 6.28}$$

A methylidyne ligand, μ_3-CH, can be introduced by oxidative addition of one C–H bond of a bridging CH_2-group as the example in Eqn 6.29 illustrates. The activation of the C–H bond leads to metal-carbon and metal-hydrogen bond formation.

$$H_2Os_3(CO)_{10} \xrightarrow[- N_2]{CH_2N_2} H_2Os_3(CO)_{10}(\mu\text{-}CH_2) \xrightarrow[-CO]{\Delta} H_3Os_3(CO)_9(\mu_3\text{-}CH)$$

Eqn 6.29

In the case of $H_3Fe_3(CO)_9CMe$, reduction of iron carbonyls is a means of generating the organic fragment and building up the cluster — for example, the reduction of $[Fe_2(CO)_8]^{2-}$ by BH_3.THF followed by acidification. The reaction of $Co_2(CO)_8$ with BH_3.THF at 258 K yields $Co(CO)_4BH_2$.THF which is an active reducing agent. Above 283 K, this compound decomposes to give $Co_4(CO)_{12}$ and $Co_3(CO)_9C(CH_2)_nOH$ (n = 4 or 5) in addition to other cobalt-containing species. The structure of $Co_3(CO)_9C(CH_2)_5OH$ is shown in Fig. 6.40, and deuterium labelling studies have confirmed that the aliphatic chain originates from the THF molecule in the precursor.

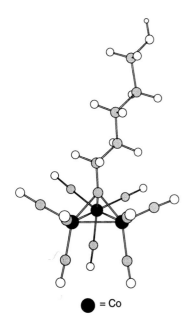

● = Co

Fig. 6.40 The structure of $Co_3(CO)_9C(CH_2)_5OH$

The reactivity of alkylidyne and related clusters has been widely studied. The halo-derivatives $Co_3(CO)_9CX$ can be derivatized as the examples in Eqns 6.30 and 6.31 illustrate, and the corresponding methylidyne cluster reacts with ethene and silanes as shown in Eqns 6.32 and 6.33. Note that in reactions 6.30 and 6.31, in addition to the intended reagent, CO has attacked the capping carbon atom.

$$Co_3(CO)_9CBr \xrightarrow{ROH} Co_3(CO)_9CCO_2R \qquad\qquad \text{Eqn 6.30}$$

$$Co_3(CO)_9CBr \xrightarrow{RNH_2} Co_3(CO)_9CC(O)NHR \qquad\qquad \text{Eqn 6.31}$$

$$Co_3(CO)_9CH \xrightarrow{CH_2=CH_2} Co_3(CO)_9CCH_2CH_3 \qquad\qquad \text{Eqn 6.32}$$

$$Co_3(CO)_9CH \xrightarrow{R_3SiH} Co_3(CO)_9CSiR_3 \qquad\qquad \text{Eqn 6.33}$$

Moving from a methylidyne to higher CR ligands opens up new possibilities for transformations of the organic group. The deprotonation of $H_3Fe_3(CO)_9CMe$ occurs with accompanying H_2 loss, leading to the formation of $[HFe_3(CO)_9C=CH_2]^-$; under an atmosphere of H_2, protonation of this anion regenerates the ethylidyne cluster (Eqn 6.34). Deuteration studies have confirmed that the proton is lost from the methyl group of $H_3Fe_3(CO)_9CMe$, and H_2 is eliminated from the metal framework.

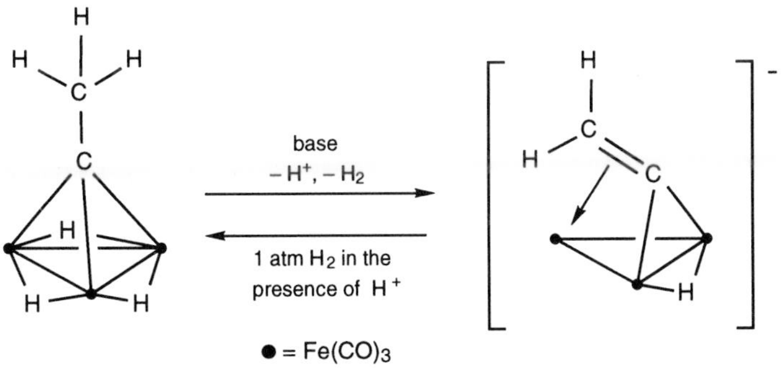

Eqn 6.34

Alkylidyne-alkyne coupling has been observed when $MeC\equiv CMe$ reacts with $H_3Ru_3(CO)_9CMe$ and the reaction is accompanied by the reduction of alkyne to alkene, specifically the Z-isomer (Eqn 6.35). The structure of the cluster product is shown in Fig. 6.41.

$$H_3Ru_3(CO)_9CMe + 2MeC\equiv CMe \rightarrow HRu_3(CO)_9(MeCCMeCMe) + Z\text{-}MeHC=CHMe$$

Eqn 6.35

Fig. 6.41 The structure of
HRu₃(CO)₉(MeCCMeCMe), and a
schematic representation that
illustrates a bonding description.

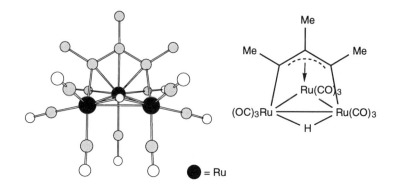

● = Ru

Similar coupling occurs between H₃Ru₃(CO)₉COMe and MeC≡CMe to
yield HRu₃(CO)₉(MeOCCMeCMe), and this compound can be
hydrogenated as shown in Eqn 6.36. The overall reaction sequence which
produces an alkylidyne unit *with an increased number of carbon atoms* is a
model for Fischer-Tropsch chain growth. Such conversions do not always
occur however, and hydrogenation of HRu₃(CO)₉(HCCHCMe) produces
H₄Ru₄(CO)₁₂.

MeC≡CMe
– H₂

H₂, 1-4 atm, 360 K
– MeOH

● = Ru(CO)₃

Eqn 6.36

6.6   Carbide clusters

We have already alluded to clusters containing interstitial atoms, and in this
section we discuss one group of these compounds: the *carbides* — clusters
containing one or more *carbon atoms*. Carbide clusters generally feature one
of the four structural motifs shown in Fig. 6.42. The carbon atom may be
fully encapsulated within the metal cage (interstitial) or may be partly
exposed. The μ₄-C atom held within the M₄-butterfly framework is often
called a semi-interstitial atom and is more reactive than either the μ₆- or μ₅-
carbon centres.

In terms of bonding schemes, whether we use a Wade approach or a total
valence electron count, an interstitial or semi-interstitial carbon atom
contributes *all four of its valence electrons* to the cluster. The total electron
count for [Rh₆(CO)₁₅C]²⁻ is 90, consistent with a trigonal prismatic metal

framework, and the total electron count for $[Fe_4(CO)_{12}C]^{2-}$ is 62 which is characteristic of a butterfly cluster. The bonding mode of the carbon atom follows from the metal geometry; in $[Rh_6(CO)_{15}C]^{2-}$ the carbon is in a μ_6-mode at the centre of the metal prism, and in $[Fe_4(CO)_{12}C]^{2-}$, the metal skeletal geometry indicates the presence of a μ_4-carbide.

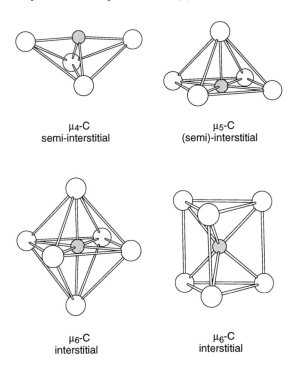

μ_4-C
semi-interstitial

μ_5-C
(semi)-interstitial

μ_6-C
interstitial

μ_6-C
interstitial

Fig. 6.42 Carbide clusters usually feature one of four structural units: the M_4-butterfly containing a μ_4-C atom, the M_5-square-based pyramid with a μ_5-C, or the M_6-octahedron or trigonal prism containing a μ_6-carbon atom.
The μ_4-carbon centre is often called semi-interstitial, and the μ_6-C is fully interstitial. The μ_5-carbon atom is normally referred to as being interstitial, but it is partly exposed and so may also be reasonably described as semi-interstitial.

See margin note on page 91 for carbide ^{13}C NMR chemical shift range.

Carbide cluster chemistry began when $Fe_5(CO)_{15}C$ was isolated in low yield from the reaction of $Fe_3(CO)_{12}$ and $MeC{\equiv}CPh$. The reaction of $Fe_5(CO)_{15}C$ with base yields $[Fe_5(CO)_{14}C]^{2-}$ which is a valuable precursor for a range of heterometallic octahedral carbide clusters (e.g. Eqns 6.37-6.39) as well as $[Fe_6(CO)_{16}C]^{2-}$ (Eqn 6.40).

$$[Fe_5(CO)_{14}C]^{2-} \xrightarrow{\ Mo(CO)_3(THF)_3\ } [Fe_5Mo(CO)_{17}C]^{2-} \qquad \text{Eqn 6.37}$$

$$[Fe_5(CO)_{14}C]^{2-} \xrightarrow{\ W(CO)_3(NCMe)_3\ } [Fe_5W(CO)_{17}C]^{2-} \qquad \text{Eqn 6.38}$$

$$[Fe_5(CO)_{14}C]^{2-} \xrightarrow{\ Rh_2(CO)_4Cl_2\ } [Fe_5Rh(CO)_{16}C]^{-} \qquad \text{Eqn 6.39}$$

$$[Fe_5(CO)_{14}C]^{2-} \xrightarrow{\ Fe_2(CO)_9\ } [Fe_6(CO)_{16}C]^{2-} \qquad \text{Eqn 6.40}$$

The formation of carbides from carbonyl clusters under conditions of pyrolysis is quite often observed, and in some reactions, there is evidence that the carbide atom originates from a CO ligand. The vacuum pyrolysis of $Ru_3(CO)_{12}$ produces octahedral $Ru_6(CO)_{17}C$ in 65% yield, and salts of

[$H_2Re_6(CO)_{18}C$]$^{2-}$, [$Re_7(CO)_{21}C$]$^{3-}$ and [$Re_8(CO)_{24}C$]$^{2-}$ can be obtained by pyrolysing a solution containing [$H_2Re(CO)_4$]$^-$. The structure of a related anion [$Re_7(CO)_{22}C$]$^-$ is shown in Fig. 6.43. The capped octahedral Re_7-core is consistent with the total valence electron count of 98, and the C atom resides in the octahedral cavity. The tetrahedral hole is too small to accommodate the carbon atom.

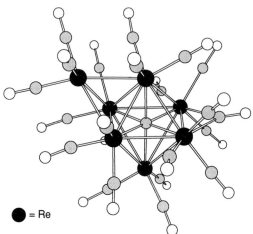

Fig. 6.43 The solid state structure of [$Re_7(CO)_{22}C$]$^-$ determined crystallographically for the tetraethylammonium salt. The Re_7-core is a monocapped octahedron, and the carbon atom is in a μ_6-mode.

● = Re

Total electron count:

7 Re [Xe]$6s^2 5d^5$ =		49
22 CO =		44
carbide =		4
1– charge =		1
Total electron count =		98

Electron count required by a monocapped octahedron = 86 + 60 – 48 = 98

The reaction of $Ru_3(CO)_{12}$ with sodium in THF gives the sodium salt of [$Ru_6(CO)_{18}$]$^{2-}$ but when heated in diglyme, this converts to the octahedral carbide cluster [$Ru_6(CO)_{16}C$]$^{2-}$ (Eqn 6.41). If the pyrolysis is continued, this time in tetraglyme, the dicarbide cluster [$Rh_{10}(CO)_{24}C_2$]$^{2-}$ is obtained. The structure of this anion is shown in Fig. 6.44 and consists of edge-sharing octahedra.

$$Ru_3(CO)_{12} \xrightarrow{\text{Na, THF, } \Delta} [Ru_6(CO)_{18}]^{2-} \xrightarrow{\text{diglyme, } \Delta} [Ru_6(CO)_{16}C]^{2-}$$

Eqn 6.41

Osmium carbide clusters can be prepared from $Os_3(CO)_{12}$ by pyrolysis in suitable ether solvents, but when [$Os_6(CO)_{18}$]$^{2-}$ is heated in triglyme, [$Os_6(CO)_{16}C$]$^{2-}$ can only be isolated in very low yield and [$Os_{10}(CO)_{24}C$]$^{2-}$ is the major product. This monocarbide anion has a tetracapped octahedral geometry (Fig. 6.45). The structural differences between the metal skeletons of [$Rh_{10}(CO)_{24}C_2$]$^{2-}$ and [$Os_{10}(CO)_{24}C$]$^{2-}$ can be rationalized in terms of electron counts.

The μ_6-carbon atom is well protected from chemical attack and much of the reactivity of these carbide clusters lies in transformations at the metal carbonyl framework rather than at the carbon centre. For example, [$Os_{10}(CO)_{24}C$]$^{2-}$ can be protonated to give $H_2Os_{10}(CO)_{24}C$ and reacts with I_2 to give [$Os_{10}(CO)_{24}(I)C$]$^-$ in which the iodine atom bridges one Os–Os edge. On the other hand, octahedral M_6C clusters may be degraded to M_5C-species as is exemplified in Eqns. 6.42 and 6.43.

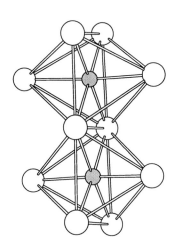

Fig. 6.44 The structure of the carbon-containing metal core of [$Rh_{10}(CO)_{24}C_2$]$^{2-}$. The structure was determined by X-ray diffraction for the [$Ph_3PCH_2CH_2PPh_3$]$^{2+}$ salt.

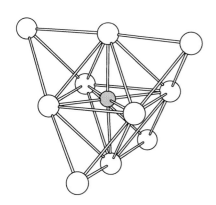

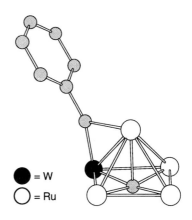

● = W
○ = Ru

Fig. 6.45 The structure of the $Os_{10}C$-core of $[Os_{10}(CO)_{24}C]^{2-}$, determined for the $[(Ph_3P)_2N]^+$ salt. The μ_6-carbide resides at the centre of an Os_6-octahedron, four faces of which are capped.

Fig. 6.46 The structure of the core of $(\eta^5\text{-Cp})WRu_4(CO)_{12}(\mu\text{-CPh})C$.

$$Ru_6(CO)_{17}C \xrightarrow[- Ru(CO)_5]{CO,\ 80\ atm,\ 340\ K} Ru_5(CO)_{15}C \qquad \textbf{Eqn 6.42}$$

$$[Fe_6(CO)_{16}C]^{2-} \xrightarrow[\text{Fe(III) or } [C_7H_7]^+]{\text{oxidation with}} Fe_5(CO)_{15}C \qquad \textbf{Eqn 6.43}$$

Each pentametal product of the above reactions possesses a square-based pyramidal cage containing a μ_5-carbide (Fig. 6.42). Carbides may also arise from scission of acetylide ligands and this has been demonstrated in the reaction of $(\eta^5\text{-Cp})WRu_2(CO)_8C_2Ph$ (with a structure similar to that shown in Fig. 6.26) with $Ru_3(CO)_{12}$ in heptane at reflux. The structure of the major product is depicted in Fig. 6.46 and reveals that the CCPh ligand in the precursor has been cleaved to give an edge-bridging CPh ligand and a μ_5-C atom. The second product of the reaction is the octahedral cluster $(\eta^5\text{-Cp})WRu_5(CO)_{14}(\mu\text{-CPh})(\mu_6\text{-C})$.

Amongst the families of carbide clusters, it is probably that of the μ_4-carbides that has played a pivotal role in the development of the area. As we saw at the beginning of the chapter, there is analogy between the M_4C-framework and stepped sites on metal surfaces, and organic transformations taking place at the μ_4-carbide have therefore been of interest. Upon treatment with bromide ion, $Fe_5(CO)_{15}C$ is degraded to $[Fe_4(CO)_{12}C]^{2-}$ (Fig. 6.47). Oxidative degradation of $[Fe_6(CO)_{16}C]^{2-}$ by iron(III) chloride in MeOH in the presence of bromide ion yields $[Fe_4(CO)_{12}CC(O)OMe]^-$ which undergoes C–C bond cleavage when treated with $BH_3.THF$ (Eqn 6.44). Proton loss from the carbide cluster so-formed gives $[Fe_4(CO)_{12}C]^{2-}$ which can be oxidized in the presence of CO to $Fe_4(CO)_{13}C$. The reactivity of the exposed carbide centre in the latter is evidenced in a series of reactions including those shown in Fig. 6.48.

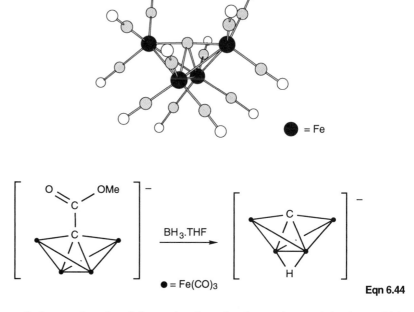

Fig. 6.47 The structure of $[Fe_4(CO)_{12}C]^{2-}$, determined for the $[PhCH_2NMe_3]^+$ salt. The μ_4-carbide lies within an Fe_4-butterfly framework but is relatively exposed.

● = Fe

● = Fe(CO)₃

Eqn 6.44

Carbon-carbon bond formation has also been observed in the carbide-alkyne coupling reaction of $H_2Ru_4(CO)_{12}C$ with $PhC{\equiv}CPh$ which occurs in hexane at 330 K to give $HRu_4(CO)_{12}CC(Ph)CHPh$. The coupling is accompanied by the transfer of one cluster-bound hydrogen atom to the organic molecule which converts the alkyne into an alkene-like ligand (Fig. 6.49).

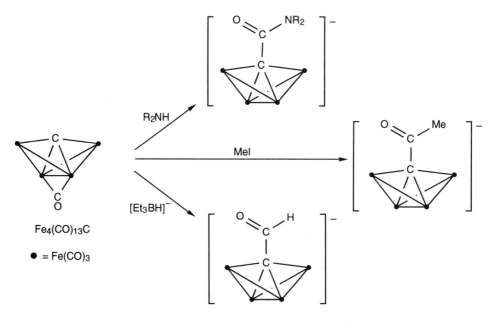

Fe₄(CO)₁₃C

● = Fe(CO)₃

Fig. 6.48 Selected reactions of Fe₄(CO)₁₃C which demonstrate the reactivity of the exposed carbide centre.

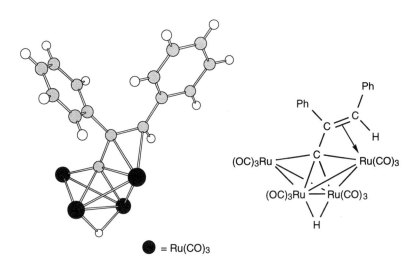

Total electron count:

4 Ru [Kr]$5s^2 4d^6$ =	32
12 CO =	24
μ-H =	1
carbide =	3
alkene π-bond =	2
Total electron count =	62

The carbide centre provides 3 (not 4) electrons because one electron is used in the localized C–C bond which is *outside* to cluster.

● = Ru(CO)$_3$

Fig 6.49   The structure of HRu$_4$(CO)$_{12}$CC(Ph)CHPh; CO ligands are omitted. The schematic representation of the cluster shows that the alkene-like organic fragment can be considered as forming one σ-C–C bond (to the μ$_4$-C) and one π-interaction.

For further discussion of the chemistry of carbide clusters, see *Further Reading* (page 4).

7 Clusters containing *p*-block atoms other than carbon

In this final chapter we consider some clusters incorporating *p*-block atoms other than carbon into the skeletal framework. In chapter 5, we described clusters with terminal phosphine (PR_3) and bridging phosphido (PR_2) ligands, and now we look at compounds in which PR groups or P atoms are intimate parts of the cluster. The discussion is further extended to cover some nitrogen- and boron-containing clusters. The discussion is necessarily selective, but further details can be found amongst the *Further reading* list (page 4).

7.1 Phosphorus

A bridging μ-PR_2 unit provides three electrons for cluster bonding.

A capping μ_3-PR unit provides four electrons for cluster bonding.

An interstitial P atom provides five electrons for cluster bonding.

On page 48 we considered how the oxidative addition of the P–H bond of a terminal PR_2H ligand can lead to the formation of an edge-bridging phosphido-ligand. This strategy for 'closing-down' the cluster framework can be extended by increasing the number of active hydrogen atoms and Eqn 7.1 shows the conversion of a μ-PPhH into a μ_3-PPh group.

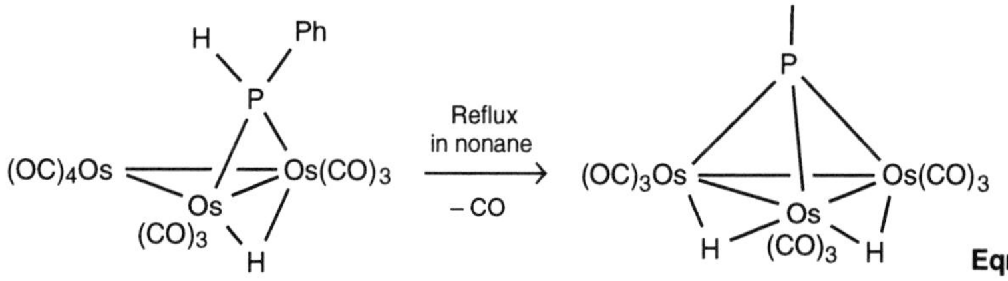

Eqn 7.1

The structures of $HOs_3(CO)_{10}(\mu\text{-PPhH})$ and $H_2Os_3(CO)_9PPh$ are shown in Fig. 7.1 along with that of the cyclometallated complex $H_2Os_3(CO)_9PPh(C_6H_4)$ which is formed from $HOs_3(CO)_{10}(\mu\text{-PPh}_2)$ but by the oxidative addition of a C_{Ph}–H bond. This unusual compound reacts with H_2 and eliminates benzene (Eqn 7.2).

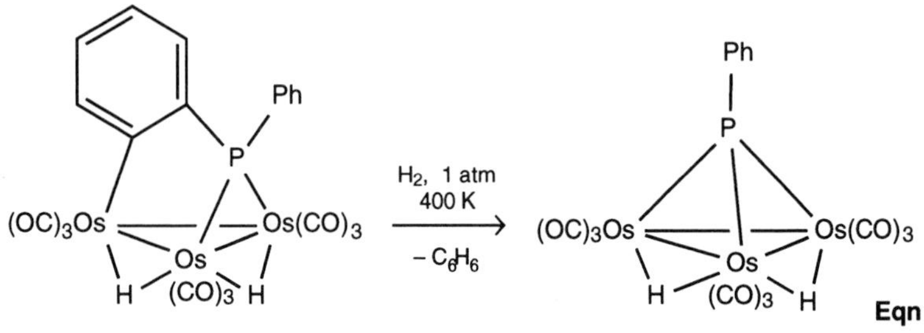

Eqn 7.2

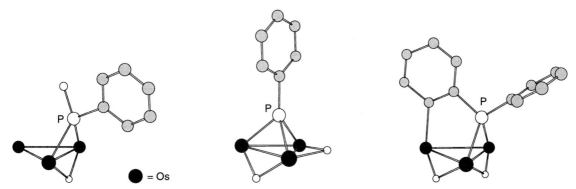

Fig. 7.1 The solid state structures, determined by X-ray diffraction methods, of HOs$_3$(CO)$_{10}$(μ-PPhH); H$_2$Os$_3$(CO)$_9$PPh; and H$_2$Os$_3$(CO)$_9$PPh(C$_6$H$_4$). CO ligands and H atoms excepting Os–H–Os bridges are omitted.

The use of phosphine, PH$_3$, permits the successive activation of three P–H bonds and the eventual formation of an interstitial phosphorus atom. The synthetic strategy is shown in Fig. 7.2 — notice that the phosphorus atom effectively pulls together two Os$_3$-triangles into the final trigonal prismatic cage. Why not an octahedron? The interstitial hole in an octahedral Os$_6$ cage is too small to accommodate the second row *p*-block atom. Although Fig. 7.2 illustrates an approach that generates the desired product in a logical progression, many cluster-forming reactions proceed in a non-specific manner. For example, the reaction of PPhH$_2$ with Ru$_3$(CO)$_{12}$ in toluene at reflux yields HRu$_3$(CO)$_{10}$(μ-PPhH), H$_2$Ru$_3$(CO)$_9$(μ_3-PPh), Ru$_4$(CO)$_{11}$(μ_4-PPh)$_2$ and Ru$_5$(CO)$_{15}$(μ_4-PPh). The structures of the last two clusters are shown in Fig. 7.3, and illustrate that capping PR groups may span four metal atoms.

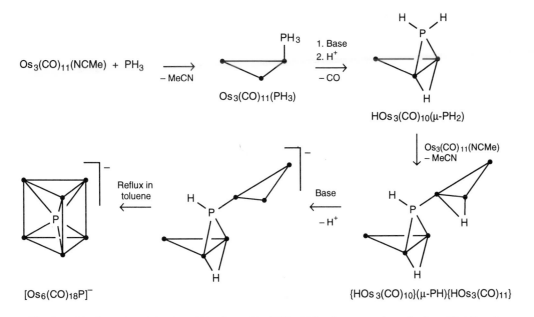

Fig. 7.2 The formation of the phosphide-cluster [Os$_6$(CO)$_{18}$P]$^-$ by the successive activation of P–H bonds.

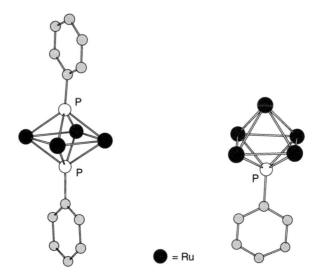

Fig. 7.3   The structures of
$Ru_4(CO)_{11}(\mu_4\text{-}PPh)_2$ and
$Ru_5(CO)_{15}(\mu_4\text{-}PPh)$ with CO ligands
and H atoms omitted.

Interstitial P atoms have been observed in trigonal prismatic (e.g. $[Os_6(CO)_{18}P]^-$) and square-antiprismatic cavities. The latter is exemplified by $[Rh_9(CO)_{21}P]^{2-}$ and $[Rh_{10}(CO)_{22}P]^{3-}$ which are prepared according to the scheme shown in Fig. 7.4.

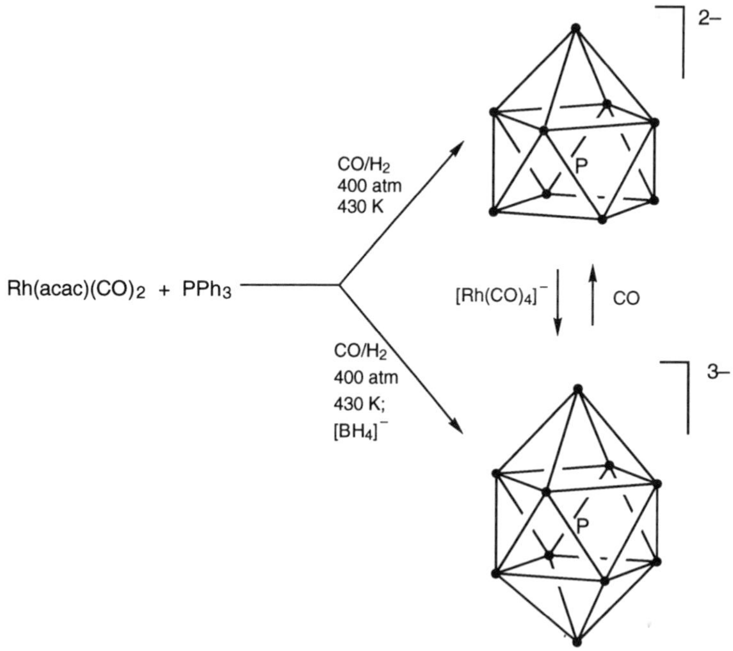

Fig. 7.4   The syntheses of
$[Rh_9(CO)_{21}P]^{2-}$ and
$[Rh_{10}(CO)_{22}P]^{3-}$, and their
interconversion. In each anion, the P
atom is interstitial and is considered to
be within bonding contact with eight
rhodium centres; the Rh_{apex}–P
distances are 306 pm compared to
243 pm for the Rh–P distances within
the square antiprismatic cavities.

Rh(acac)(CO)$_2$ + PPh$_3$

CO/H$_2$
400 atm
430 K

$[Rh(CO)_4]^-$ CO

CO/H$_2$
400 atm
430 K;
$[BH_4]^-$

Besides occurring in interstitial environments, a phosphorus *atom* (as distinct from a PR group) has been observed capping a Co$_3$-triangle in $Co_3(CO)_9P$. This cluster is one of several products in the reaction of $Co_2(CO)_8$ and white phosphorus (Eqn 7.3), or can be formed by treating

$[Co(CO)_4]^-$ with phosphorus(III) halides. Each product in Eqn 7.3 is related to tetrahedral P_4 by the replacement of a P atom by a $Co(CO)_3$ fragment. The trimerization of $Co_3(CO)_9P$ occurs spontaneously (Eqn 7.4) and the structure of the cyclic product is proposed to be analogous to that of $\{Co_3(CO)_8As\}_3$ shown in Fig. 7.5. The formation of these trimers illustrates that the capping P (or As) atom can function as a Lewis base, and further evidence of this behaviour comes from the reaction of $Co_3(CO)_9P$ with $Fe_2(CO)_9$ (Eqn 7.5 and Fig. 7.6). As the Lewis basicity of the capping atom in the clusters $Co_3(CO)_9E$ (E = P, As, Sb, Bi) decreases in descending group 15, the tendency for trimerization also decreases; the monomer $Co_3(CO)_9Bi$ is stable with respect to oligomerization.

$$Co_2(CO)_8 + P_4 \xrightarrow{\text{CO, hexane}} Co(CO)_3P_3 + Co_2(CO)_6P_2 + Co_3(CO)_9P \qquad \textbf{Eqn 7.3}$$

$$3Co_3(CO)_9P \xrightarrow{\text{on standing in hexane}} \{Co_3(CO)_8P\}_3 + 3CO \qquad \textbf{Eqn 7.4}$$

$$Co_3(CO)_9P \xrightarrow{\text{Fe}_2(CO)_9,\ \text{THF}} Co_3(CO)_9P\{Fe(CO)_4\} \qquad \textbf{Eqn 7.5}$$

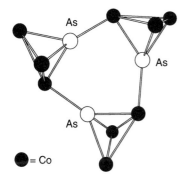

Fig. 7.5 The structure of the trimer $\{Co_3(CO)_8As\}_3$.

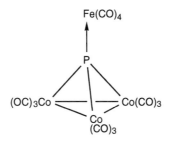

Fig. 7.6 The $Co_3(CO)_9P$ cluster can act as a Lewis base formally replacing CO in $Fe(CO)_5$.

7.2 Nitrogen

In organometallic clusters, the most commonly observed nitrogen-containing units (excluding terminally coordinated ligands) are μ_3-NR (R = H, alkyl or aryl), μ_4-N, μ_5-N and μ_6-N. The last three modes of bonding are directly analogous to the carbides shown in Fig. 6.42. Molecular clusters in which a metallic framework supports NH_x-units may be considered as possible models for metal surface species involved in, for example, the Haber process, and cluster reactions that have been studied include those which show N–H bond cleavage and N–N bond formation and cleavage. In the synthesis of M_xNR clusters, use is often made of coordinated nitrosyl (NO) ligands (Fig. 7.7). The transformations involving the metal-bound NO group have some relevance to studies of atmospheric pollution by nitric oxide.

A bridging μ-NR_2 unit provides three electrons for cluster bonding.

A capping μ_3-NR unit provides four electrons for cluster bonding.

An interstitial N atom provides five electrons for cluster bonding.

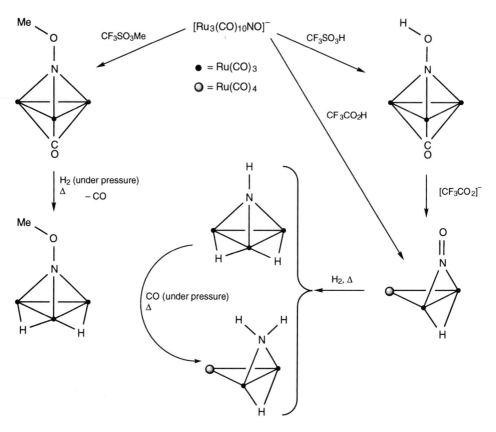

Fig. 7.7   Selected examples, taken from triruthenium chemistry, of the use of a metal-bound nitrosyl ligand in the formation of μ_3-NR groups. Notice how important the choice of acid is for the protonation of $[Ru_3(CO)_{10}NO]^-$.

Several methods of introducing μ_3-NAr (Ar = aryl) groups into triruthenium clusters are shown in Fig. 7.8. The competition between the formation of an Ru_3NAr- or $Ru_3(NAr)_2$-framework in these reactions is highly dependent upon temperature, solvent and the aryl ring substituent. In the right-hand reaction, the product distribution depends on temperature and solvent; e.g. for X = H, raising the temperature to the boiling point of THF or using hexane as solvent swings the reaction in favour of $Ru_3(CO)_9(\mu_3$-$NPh)_2$.

The nitrosyl ligand also features in a range of reactions that result in the formation of *nitrido*-clusters containing a μ_4-nitrogen atom within a butterfly framework of metal atoms. Selected reactions are given in Fig. 7.9. In comparison with the μ_4-carbide centre (Fig. 6.3), attempted protonation of the μ_4-N atom in $[Ru_4(CO)_{12}N]^-$ results in protonation of the metal framework, although there is evidence that initial attack by the proton is indeed at the nitrogen atom (Fig. 7.9). The conversion of the tetrahedral $[H_3Ru_4(CO)_{12}]^-$ into the nitrido-cluster $HRu_4(CO)_{12}N$ can be viewed in terms of the initial insertion of NO^+ into an Ru–Ru bond. Further evidence comes from the analogous reaction of $[H_3Os_4(CO)_{12}]^-$ with NO^+ in which the major product is $H_3Os_4(CO)_{12}(\mu$-NO).

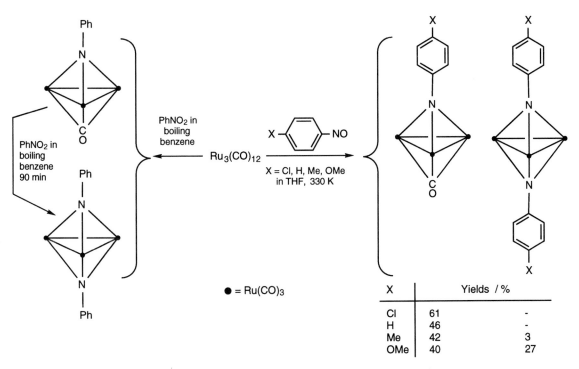

Fig. 7.8   Methods of incorporating μ3-NAr groups into triruthenium clusters.

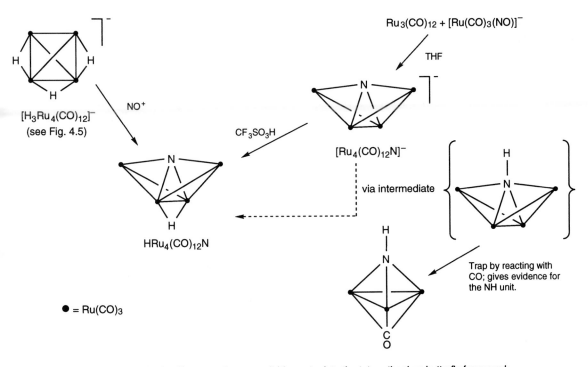

Fig. 7.9   Methods of incorporating a μ4-nitrido centre into the tetraruthenium butterfly framework.

Reactions of the μ_4-N atom have not been as extensively explored as those of the corresponding carbide systems, but when diphenylacetylene reacts with [Ru$_4$(CO)$_{12}$N]$^-$, the alkyne inserts into an Ru–Ru bond rather than interacting with the nitrogen centre.

Routes to clusters containing a μ_5-N atom may be similar to those to the μ_4-nitrides. Equations 7.6 and 7.7 compare the conditions under which [Fe$_4$(CO)$_{12}$N]$^-$ and [Fe$_5$(CO)$_{14}$N]$^-$ form from [Fe$_2$(CO)$_8$]$^{2-}$. Alternatively, a cluster expansion reaction can be used to increase the degree of encapsulation of the N atom (Eqn 7.9). The M$_5$-nitrides possess square-pyramidal skeletons, consistent with an electron count of 74.

$$[Fe_2(CO)_8]^{2-} + [NO]^+ \xrightarrow{400\ K} [Fe_4(CO)_{12}N]^- \qquad \textbf{Eqn 7.7}$$

$$[Fe_2(CO)_8]^{2-} + [NO]^+ \xrightarrow{>\,400\ K} [Fe_5(CO)_{14}N]^- \qquad \textbf{Eqn 7.8}$$

$$[FeRu_3(CO)_{12}N]^- + Ru_3(CO)_{12} \xrightarrow{Reflux\ in\ THF} [FeRu_4(CO)_{14}N]^- \qquad \textbf{Eqn 7.9}$$

Interstitial μ_6-N atoms in both octahedral and trigonal prismatic cages are known and include [Co$_6$(CO)$_{13}$N]$^-$ (octahedral), [Co$_6$(CO)$_{15}$N]$^-$ (trigonal prismatic) and [Ru$_6$(CO)$_{16}$N]$^-$ (octahedral). The cobalt clusters are prepared according to the sequence in Eqn. 7.10, and the collapse of the trigonal prismatic to octahedral cluster is consistent with a change in the total cluster electron count from 90 to 86 as two CO ligands are expelled. This reaction illustrates the stabilizing effect that an interstitial atom may have — the metal skeleton can reorganize itself with the *p*-block atom essentially holding the metal fragments in place.

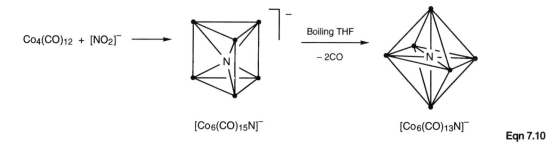

$$\text{Co}_4(\text{CO})_{12} + [\text{NO}_2]^- \longrightarrow \quad [\text{Co}_6(\text{CO})_{15}\text{N}]^- \xrightarrow[-\,2\text{CO}]{Boiling\ THF} [\text{Co}_6(\text{CO})_{13}\text{N}]^-$$

Eqn 7.10

7.3   Boron

After reading sections 6.4, 6.5 and 7.2, the reader should be aware of structural similarities between some carbon and nitrogen-containing clusters. For example the pair of compounds H$_3$Ru$_3$(CO)$_9$CH and H$_2$Ru$_3$(CO)$_9$NH differ in that a {C+H}-combination has been formally replaced by an isoelectronic N atom. Structurally, the compounds are quite similar. On moving to boron, the series of compounds may, on paper at least, be extended to 'H$_4$Ru$_3$(CO)$_9$BH' which possesses the same number of cluster bonding electrons as each of H$_3$Ru$_3$(CO)$_9$CH and H$_2$Ru$_3$(CO)$_9$NH. In practice, the metallaborane can be prepared by treating $Ru_3(CO)_{12}$ with $BH_3.THF$ and $Li[BHEt_3]$ followed by protonation.

In solution, $Ru_3(CO)_9BH_5$ exists as two isomers (Fig. 7.10) and the bridging hydrogen atoms are involved in a fluxional process which results in the interconversion of the two structures. Historically, the first compound of this type to be isolated and characterized was the triiron analogue, and in solution and the solid state, the hydrogen distribution in $Fe_3(CO)_9BH_5$ is that found for isomer **A** of $Ru_3(CO)_9BH_5$. The borane fragment in $Fe_3(CO)_9BH_5$ is susceptible to attack by Lewis bases, L, and can be removed from the cluster in the form of $L.BH_3$. The presence of the relatively high number of hydrogen atoms on the surface of these clusters can be utilized in synthetic pathways to higher nuclearity clusters. For example, the loss of H_2 from $Fe_3(CO)_9BH_5$ is a driving force for cluster expansion to $HFe_4(CO)_{12}BH_2$ (Eqn 7.11). The structure of this tetrairon cluster is shown in Fig. 7.11, and is similar to the butterfly carbides and nitrides that we have described earlier. However, the skeleton must accommodate additional H atoms to maintain the required total electron count of 62.

$$Fe_3(CO)_9BH_5 + 2Fe_2(CO)_9 \rightarrow HFe_4(CO)_{12}BH_2 + 3Fe(CO)_5 + H_2 \qquad \textbf{Eqn 7.11}$$

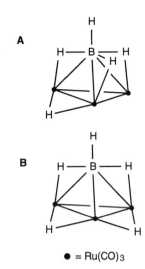

Fig. 7.10 The two isomers of $Ru_3(CO)_9BH_5$ which are observed in solution.

Fig. 7.11 The solid state structure of $HFe_4(CO)_{12}BH_2$.

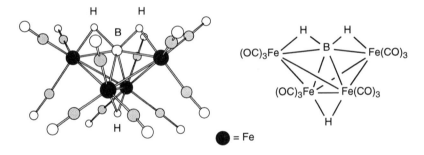

Total electron count :

4 Fe $[Ar]4s^2 3d^6 =$	32
12 CO =	24
boride=	3
3 H atoms =	3
Total electron count =	62

On standing in solution, or when photolysed, $Ru_3(CO)_9BH_5$ undergoes a self-expansion reaction to form a mixture of $HRu_4(CO)_{12}BH_2$ (which is structurally analogous to $HFe_4(CO)_{12}BH_2$) and $HRu_6(CO)_{17}B$ (Fig. 7.12).

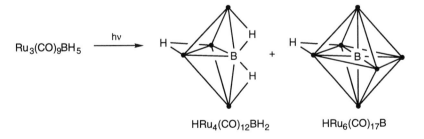

Fig. 7.12 Dihydrogen loss is a driving force in the self-expansion of $Ru_3(CO)_9BH_5$ to boride clusters. The products have been drawn so as to emphasize the structural relationship between them.

Like the μ_4-C in the butterfly carbides, the μ_4-boron atom is a relatively reactive centre. Unlike the carbide however, the boron atom is associated with *two* cluster-bound H atoms, and deprotonation is often an initial step in reactions at this site. Interestingly, the order in which protons are lost is different in the iron butterfly to that in the ruthenium and osmium clusters — for iron, the sequence of loss is Fe-*H*-B, then Fe-*H*-Fe, then Fe-*H*-B, but for ruthenium and osmium two protons are lost sequentially from M-*H*-B sites.

The anions $[HRu_4(CO)_{12}B]^-$ and $[HFe_4(CO)_{12}B]^-$ undergo metal-cage expansions as illustrated for ruthenium in Fig. 7.13. This method gives a route to heterometallic cages containing interstitial boron. In the example chosen, both *cis*- and *trans*-isomers of the mixed rhodium-ruthenium cluster are produced in the reaction, but the *trans*-isomer is favoured.

The coupling of a carbide and an alkyne was illustrated in Fig. 6.49, but we noted in section 7.2 that related nitrido-compounds did not undergo similar reactions undergoes coupling to alkynes. Irradiation of a solution containing $HRu_4(CO)_{12}BH_2$ and PhC≡CPh leads to the formation of $HRu_4(CO)_{12}BHC(Ph)CPhH$ (Fig. 7.13), and structural data for this compound confirm that B–C bond formation occurs. A comparison of Figs. 7.13 and 6.49 show that the carbide- and boride-coupled products are significantly different; in the former, the Ru_4-butterfly framework stays intact, but in the latter, one Ru–Ru edge is broken. Such a difference is hard to rationalize.

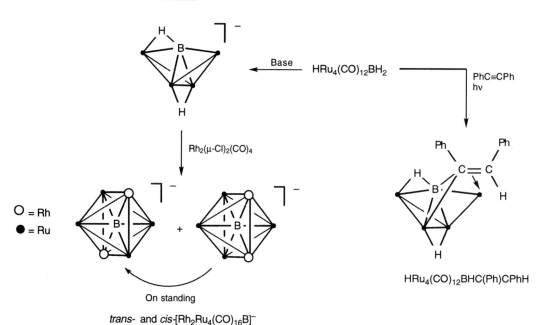

Fig. 7.13　Selected reactions of $HRu_4(CO)_{12}BH_2$, showing examples of metal-cage expansion around the boron atom, and boron-carbon coupling.

Problems

The first section contains problems concerning structure and bonding and the answers lie within the text, or are closely related to discussions in the text. Problems in the next section are taken from real experimental situations; the answers can be found by consulting the reference cited but *a word of encouragement*: there is not always a unique solution to the spectroscopic data and if your answer does not match that in the cited paper, consider it carefully. It may be an alternative solution consistent with the data given in the question. What other experimental data are needed to confirm your (or the literature) structures?

Section 1: Structure and bonding

1. The following pairs of fragments are isolobal:
   CH_3 and $Mn(CO)_5$ CH_2 and $Os(CO)_4$ $[CH]^+$ and $Fe(CO)_3$
   What do you understand by this statement and what is implied about the geometry of each fragment?

2. Fig. 2.1 (page 5) shows the structure of $Mn_2(CO)_{10}$ in which each Mn centre obeys the 18-electron rule. What other structures would be consistent with 18-electron Mn centres? Are any of these structures reasonable in practice? On what do you base your decisions?

3. Show that each metal atom is an 18-electron centre in each of the following compounds:

$Fe_3(CO)_{12}$	(Fig. 2.10, page 12)
$Rh_4(CO)_{12}$	(Fig. 2.16, page 15)
$(\eta^5\text{-Cp})_2Fe_2(CO)_4$	(Fig. 6.4, page 56)
$(\eta^5\text{-Cp})_2Ni_2(\mu\text{-PPh}_2)_2$	(Fig. 6.10, page 59)
$(\eta^5\text{-Cp})_2Mo_2(CO)_4$	(Fig. 6.13, page 60)

   What limitations does this localized bonding scheme have in cluster compounds?

4. Predict the cluster core geometries of the following:

$Os_4(CO)_{16}$	$[Os_6(CO)_{18}P]^-$
$[H_3Os_4(CO)_{12}]^-$	$[Co_6(CO)_{13}N]^-$
$Os_5(CO)_{16}$	$Ru_5(CO)_{15}(PPh)$
$H_2Os_5(CO)_{16}$	$H_3Os_3(CO)_9CMe$
$Os_7(CO)_{21}$	$[Os_4(CO)_{12}N]^-$
$[CoFe_2Ru(CO)_{13}]^-$	$Os_6(CO)_{18}$

   Would you expect $[Mn_3(CO)_{14}]^-$ to possess a triangular framework?

5. Describe the various sites and bonding modes that hydride ligands can occupy in dimetal and metal-cluster compounds. How can solution ^1H NMR spectroscopic data provide information about the positions of hydride ligands? Why may the conclusions reached from solution studies be different from data obtained in the solid state?

6. Confirm that each of the following condensed polyhedral structures is consistent with the number of valence electrons available.

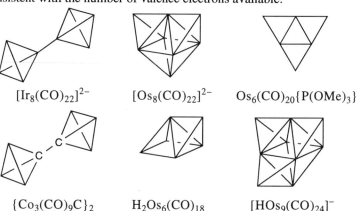

$[Ir_8(CO)_{22}]^{2-}$ $[Os_8(CO)_{22}]^{2-}$ $Os_6(CO)_{20}\{P(OMe)_3\}$

$\{Co_3(CO)_9C\}_2$ $H_2Os_6(CO)_{18}$ $[HOs_9(CO)_{24}]^-$

Nuclear spin data required for Section 2

Nucleus	Spin	%Abundance
^1H	$^1/_2$	>99.9
^{13}C	$^1/_2$	1.1
^{31}P	$^1/_2$	100
^{103}Rh	$^1/_2$	100

Abbreviations in NMR spectroscopic data:

s	singlet
m	multiplet
t	triplet
d	doublet
dd	doublet of doublets
ddd	doublet of doublets of doublets

Relative atomic masses are given in the periodic table on the inside back-cover of this book.

Section 2: Interpretation of experimental data

1. The reaction of $(\eta^5\text{-}Cp)_2Cr_2(CO)_6$ with white phosphorus at 363 K leads to the formation of two cluster compounds **A** and **B**. Using the following data, suggest identities for the products, and give possible structures for them.

Compound	Parent ion in mass spectrum / amu	^{31}P NMR spectrum
A	408	One signal
B	266	One signal

[L.Y. Goh *et al.*, (1989) *J. Chem. Soc., Dalton Trans.*, 1951-1956.]

2. On heating a mixture of $(\eta^5\text{-}Cp)_2Mo_2(CO)_6$ and P_2Ph_4, compound **A** is formed. When **A** is photolysed, CO is lost and **B** is produced. Treatment of compound **A** with tetrafluoroboric acid (HBF$_4$) gives cation **C**. Use the data given below to identify the products, and suggest possible structures for them. What are the formal metal–metal bond orders in $(\eta^5\text{-}Cp)_2Mo_2(CO)_6$ and in each product?

Compound	^1H NMR spectrum	^{31}P NMR spectrum	IR spectrum / cm^{-1}
A	7.8-7.2 (m, 20H), 5.4 (s, 10H)	One signal	1855
B	7.2-6.5 (m, 20H), 5.6 (s, 10H)	One signal	1674
C	7.8-6.8 (m, 20H), 5.5 (s, 10H), −13.4 (t, *J* 56 Hz, 1H)	One signal	2010

[T. Adatia *et al.* (1989) *J. Chem. Soc., Dalton Trans.*, 1555-1564.]

3. The reaction of $Os_3(CO)_{11}(NCMe)$ with PH$_3$ followed by treatment with base and then acid gives compound **A**, the ^1H NMR spectrum of which exhibits signals at δ 5.2 (m), 4.9 (dd) and −19.9 (dd). Suggest a structure for compound **A** and interpret the NMR spectroscopic data. [See Section 7.1.]

4. $H_3Ru_3(CO)_9(\mu_3\text{-Bi})$ possesses a CO and H arrangement similar to that in $H_3Fe_3(CO)_9CMe$ (Fig. 6.39). Reaction with PPh_3 gives product **A** ($^m/_e$ 1001) for which the ^1H NMR spectrum shows signals at 7.25-7.45 (m, 15H), −17.2 (dd, *J* 10.5, 2.9 Hz, 2H) and −17.4 (dt, *J* 2.9, 1.3 Hz, 1H). What is the structure of the substituted product?
   [B.F.G. Johnson *et al.* (1988) *J. Chem. Soc., Dalton Trans.*, p. 3129-3135.]

$\eta^5\text{-Cp}^* = \eta^5\text{-C}_5Me_5$

5. The unsaturated cluster $(\mu\text{-H})_3(\mu_3\text{-H})(\eta^5\text{-Cp}^*)_3Co_3$ reacts with CO to yield a saturated cluster **A**. The IR spectrum of **A** shows two strong CO absorbances (1778 and 1652 cm^{-1}); in addition to signals due to the Cp* carbon atoms, only *one* downfield resonance is observed in the ^{13}C NMR spectrum (δ +271). The ^1H NMR spectrum of **A** has resonances at δ 1.63 (s, 15H), 1.56 (s, 30H) and −32.3. On heating **A**, H_2 is lost to give the tricobalt cluster **B**, the ^1H NMR spectrum of which shows only one signal (δ 3.4). The IR spectrum of **B** shows an absorption at 1681 cm^{-1}. Rationalize these data, and suggest possible structures for **A** and **B**.
   [C.P. Casey *et al.* (1993) *J. Chem. Soc., Chem. Commun.*, 1692-1694.]

Remember that the timescales of IR and NMR spectroscopies are different.

6. The phosphine-substituted cluster $Ir_4(CO)_{11}(PPhH_2)$ reacts with $[(\eta^5\text{-Cp}^*)Ir(NCMe)_3][SbF_6]_2$ in the presence of base (DBU) to give a product **A** with the following spectroscopic characteristics:
IR / cm^{-1}	2072, 2040, 2024 (all strong), 1855, 1832 (both weak)
Mass spec.	1483 (parent ion)
^{31}P NMR	one signal
^1H NMR δ	2.1 (d, *J* 1.6, 15H) in addition to a multiplet ≈ 7.5
   Suggest the identity of **A** and give a possible structure. Are isomers possible? [R. Khattar *et al.* (1990) *Organometallics*, **9**, 645-656.]

Approximate ^{13}C NMR spectroscopic shift ranges for carbonyl and carbide C nuclei.

carbides	μ-CO	terminal CO
▬▬	▬▬	▬▬

+400		+200 δ

7. When $H_3Ru_3(CO)_9(\mu_3\text{-CCO}_2Me)$ **A** is heated a rearrangement occurs. The IR spectrum of the product **B** shows absorptions due to terminal CO ligands plus a band at 1540 cm^{-1}; the corresponding absorption was at 1685 cm^{-1} in **A**. At 298 K, the ^1H NMR spectrum of **B** has signals at 3.9 (s, 1H), 3.5 (s, 3H), −13.0 (broad, 1H), −14.9 (broad, 1H). At 263 K, the two broad signals become doublets (*J* 3 Hz). Pyrolysis of **A** in the presence of CO does not give **B** but instead leads to $Ru_3(CO)_{12}$ and the elimination of methyl acetate. Identify **B** and explain what features of **A** facilitate the rearrangement.
   [M.R. Churchill *et al.* (1987) *Organometallics*, **6**, 799-805.]

8. Using electron counting rules and the solution spectroscopic data (173 K) below, deduce a possible structure (including CO locations) for $[HRh_6(CO)_{13}C]^-$.
^1H NMR	δ −14.6 (t, *J* 20 Hz)
^{103}Rh NMR	Equal intensity: δ −288 (s), −749 (d, *J* 20 Hz), −996 (s)
^{13}C NMR (J Rh-C given)	δ 192 (d, *J* 78 Hz, 2C), 194 (d, *J* 89 Hz, 2C), 195 (d, *J* 93 Hz, 2C), 209 (ddd, *J* 54, 18, *J*(C-H)19 Hz, 2C), 215 (t, *J* 40 Hz, 1C), 236 (pseudo triplet, *J* 39, 39 Hz, 2C), 232 (dd, *J* 43, 29 Hz, 2C), 460 (septet, *J* 17 Hz, 1C).
   [S. Bordoni *et al.* (1988) *J. Chem. Soc., Dalton Trans.*, 2103-2108.]

Index